T0223859

Lecture Notes in Artificial Intelligence 764

Subseries of Lecture Notes in Computer Science
Edited by J. G. Carbonell and J. Siekmann

Lecture Notes in Computer Science
Edited by G. Goos and J. Hartmanis

Gerd Wagner

Vivid Logic

Knowledge-Based Reasoning with Two Kinds of Negation

Springer-Verlag

Berlin Heidelberg New York
London Paris Tokyo
Hong Kong Barcelona
Budapest

Series Editors

Jaime G. Carbonell
School of Computer Science, Carnegie Mellon University
Schenley Park, Pittsburgh, PA 15213-3890, USA

Jörg Siekmann
University of Saarland
German Research Center for Artificial Intelligence (DFKI)
Stuhlsatzenhausweg 3, D-66123 Saarbrücken, Germany

Author

Gerd Wagner
Institut für Philosophie (WE1), Freie Universität Berlin
Habelschwerdter Allee 30, D-14195 Berlin, Germany

CR Subject Classification (1991): I.2.3-4, F.4.1

ISBN 3-540-57604-5 Springer-Verlag Berlin Heidelberg New York
ISBN 0-387-57604-5 Springer-Verlag New York Berlin Heidelberg

Typesetting: Camera ready by author
45/3140-543210 - Printed on acid-free paper

To the crocodiles, cockatoos and flying foxes I met in Australia
when attending IJCAI'91,

and to all other creatures of wildlife

endangered by our industrial civilization all over the world,
suffering from,
and threatened by extinction through
mass tourism, environmental pollution and industrial exploitation
of the oceans by overfishing (especially with drift nets),
and of tropical rain forests by overlogging,
with the participation of multinational companies
which, at the same time, develop fifth generation computer systems

while we are doing AI.

Preface

This work has been carried out within an interdisciplinary research project "Systeme der Logik als theoretische Grundlage der Wissens- und Informationsverarbeitung" at the Institute for Philosophy of the Free University of Berlin. I would like to express my gratitude to the Free University for giving me the opportunity to pursue my own research within a very free but supportive working environment. I thank my thesis advisers and colleagues David Pearce, Heinrich Herre, Heinrich Wansing, and Marcus Kracht for their help and inspiration in numerous conversations. I also want to thank

- Wolfgang Rautenberg, who was my teacher in what turned out to be essential for my work: *nonclassical logics*,

- Michael Gelfond for a stimulating discussion and helpful suggestions concerning the chapter on vivid knowledge representation and reasoning,

- Roland Bol for critical remarks and suggestions concerning the chapter on logic programming with strong negation,

- Jan Jaspars for discussing with me the topic of information growth and pointing out a shortcoming in my first definition of the informational extension of a logic program.

- Yao Hua Tan and Wiebe van der Hoek for inviting me to the International Workshop on Nonmonotonic Reasoning and Partial Semantics 1991 in Amsterdam,

- Patrick Doherty and Dimiter Driankov for their invitation to give a lecture at the Linköping Summer School on Nonmonotonic Reasoning and Partial Semantics in Knowledge Representation 1992,

- André Fuhrmann and Hans Rott for inviting me to the Constance Colloquium in Logic and Information 1992.

I have much benefitted from contacts and discussions with colleagues especially at the workshops on nonmonotonic reasoning and partial semantics in Amsterdam and Linköping.

Furthermore, I am grateful to Georg Schneider who helped me a lot. Special thanks also to Birgit (for many things). Congratulations to Heike who managed to finish her thesis before I did. Finally, thanks to Daniela for sharing with me the experience of crocodiles, cockatoos and IJCAI'91.

Parts of the material of this thesis have been published or will appear in the form of articles:

- A modified version of Chapter 2 has appeared in A. Fuhrmann and H. Rott (Eds.), *Logic, Action and Change*, de Gruyter, 1993.

- Chapter 5 has appeared in the *Proceedings of the First All-Berlin Workshop on Nonclassical Logics and Information Processing*, edited by D. Pearce and H. Wansing, Lecture Notes in Artificial Intelligence 619, Springer Verlag, Berlin, 1992, pp. 80–91.

- Chapter 6 has been published in the *Journal of Logic and Computation* Vol. 1, No. 6, 1991, pp. 835–859.

- Chapter 7 has appeared in two different versions as

   - A Database Needs Two Kinds of Negation, in B. Thalheim, J. Demetrovics and H.-D. Gerhardt (Eds.), *Proceedings of 3rd Symposium on Mathematical Fundamentals of Database and Knowledge Base Systems (MFDBS-91)*, Lecture Notes in Computer Science 495, Springer Verlag, Berlin, 1991, pp. 357–371.

   - Vivid Reasoning with Negative Information, in W. van der Hoek, J.-J. Ch. Meyer, Y.H. Tan and C. Witteveen (Eds.), *Non-Monotonic Reasoning and Partial Semantics*, Series in Artificial Intelligence, Ellis Horwood, 1992, pp. 181–205.

The danger in any formal logical investigation is that results may be regarded as merely formal or mathematical, with no relevance to the 'real world'. I hope that this is not the case in the research presented here, since it was the main goal to make theoretical contributions which are relevant to the design and understanding of practical knowledge representation systems.

Contents

Chapter 1

General Introduction

Reasoning as a cognitive activity is based on knowledge available to the cognitive agent through some form of representation, e.g. in memory, or on a sheet of paper, or in a computer database. The goal of knowledge representation (KR) research in computer science, therefore, is to improve the existing KR technology, such as relational database systems, in order to provide more sophisticated KR services to humans using them as extensions of their own biological ressources.[1]

While 'knowledge-based reasoning' sounds like a pleonasm (isn't reasoning always based on knowledge or belief ?), the subtitle of this book is to emphasize the difference between certain forms of commonsense reasoning (namely those processing the various kinds of information available in a knowledge base) and theory-based reasoning, such as in mathematics. In real life people make judgements and draw conclusions even if they do not have complete information about a problem, and even if the information at hand contains incompatible items. This kind of *practical reasoning* is distinct from standard deductive and probabilistic reasoning. Most philosophers and logicians,[2] however, paid more attention to *theoretical reasoning* which deals with exact (totally defined) predicates without truth-value gaps excluding the possibility of incoherent information. In this highly idealized framework there is no need for the representation of negative information: it suffices, for instance, to define the predicate *odd_integer* in a purely positive fashion in order to be able to derive also negative statements like ¬*odd_integer*(2). In many forms of commonsense reasoning, however, we also use (explicit or implicit) negative information in addition to our positive knowledge in order to conclude something, or even to 'jump' to a conclusion. It is this kind of knowledge representation and reasoning realized by means of appropriately extended database systems which I have in mind when I speak of

---

[1] Of course, the ultimate goal in Artificial Intelligence is to build autonomous intelligent systems not only capable of supporting humans, but also capable of solving cognitive problems and managing complex tasks on their own.

[2] Notable exceptions are, for instance, Toulmin [1956], Körner [1966], Rescher [1976].

'knowledge-based reasoning'.

Knowledge representation is one of the central topics in Artificial Intelligence (AI). In the beginning of its study there was hope that one could take the already well-established classical first-order predicate logic from mathematics, and use it as a formal language for knowledge representation systems. The attempt to implement it on a computer has lead to a whole enterprise on its own, called *automated theorem proving*, which is mainly guided by the need to find more efficient techniques.

While the efficiency problems in automated theorem proving based on classical predicate logic caused doubts about its usefulness for KR, even greater concerns have arisen with respect to its adequacy regarding commonsense reasoning. Many phenomena of commonsense reasoning, like nonmonotonic inferences, or the use of different kinds of negation, cannot be accounted for in classical logic which was rather designed as a model for mathematical reasoning. Therefore, a great variety of alternative logics have been, and are still, investigated by AI researchers in order to evaluate their usefulness for the automation of commonsense knowledge representation and reasoning.

For a long time the search for underlying patterns of reasoning that lay outside the scope of classical standard logic was the main concern in philosophical logic. Although work in philosophical logic mostly neglected the issues of computability and computational complexity its results are still important to AI and KR research. Besides epistemic logics, notably partial, paraconsistent and constructive logics are to be mentioned as particularly interesting theoretical frameworks because they model cognitive phenomena which are at the centre of KR.

As Donald Nute remarks in an Editorial of an issue of the Journal of Logic and Computation [Nute 1990], philosophical logic is not only formal, it is also *descriptive* (attempting to capture empirical phenomena) and *normative* (prescribing appropriate reformations of our imperfect conceptual schemes). The same holds for KR research. The aim is to establish a formal system which:

1. captures a practically relevant body of cognitive facilities employed by humans which can be reasonably implemented, and

2. extends the knowledge representation and reasoning capabilities of humans by capitalizing on the technical strength of the system including formal techniques to master 'jams' (situations where we are inclined to accept incompatible statements) and 'gaps' (situations where we want to draw a conclusion but are uncertain because of missing information) in a principled way.

Of course, in KR we have the additional requirement that any useful formalization must be able to be implemented in a computer program. Nute: "If the formalization cannot be implemented in a computer program, then this is a fairly strong piece of evidence that it does not represent the way people actually use the concepts the formalism is supposed to capture. [...] Rather than first developing a formal theory in some abstract symbolic language and then implementing that theory in a program, why should we not develop the formal theory in a programming language from the beginning ? [...] On this view, any programming effort that attempts to simulate and improve upon human reasoning using an explicit knowledge representation method is an exercise in philosophical logic."

1.1 Overview

In this book, I actually follow Nute's perspective by developing formalisms based on, and conservatively extending, computational systems like databases and logic programs. After formulating the concept of a *knowledge representation and reasoning system (KRS)* as a general framework, I define the notion of a *vivid* KRS (VKRS)[3] which is characterized by the presence of two kinds of negation, and the requirements of Restricted Reflexivity, Constructivity and Non-Explosiveness.

The concept of a KRS consists essentially of two main components: an answer and an update operation manipulating knowledge bases as abstract objects,[4] together with a set of formal properties these operations may have. In general, there are no specific restrictions on the internal structure of a knowledge base. We will see, however, that a computational design can be achieved by 'compiling' incoming information into some normal form rather than leaving it in the form of arbitrarily complex formulas. This is the case, for instance, in Belnap's KRS which can be considered as the paradigm for KRSs.

The concept of a KRS constitutes a useful framework for the classification and comparison of various AI systems and formalisms. It is more general than that of a logic (i.e. a consequence relation). A standard logic can be viewed as a special kind of KRS. On the other hand, by defining the answer and update operations procedurally, KRSs can serve as the basis for the operational definition of logics. There is even a strong analogy between the concept of a KRS and that of a Gentzen sequent system.

The concept of a VKRS is a two-fold generalization:

1. it extends already known logics, such as Belnap's 4-valued or Nelson's paraconsistent constructive logic, by adding *weak* negation, and

2. it extends already known knowledge representation systems, such as relational or deductive database systems, by adding *strong* negation.

---

[3]The idea of *vividness* as a design principle for KR systems has first been suggested by Levesque [1986].

[4]This with respect to KR fundamental distinction was first proposed in [Levesque 1984a] where the operations are called *ASK* and *TELL*.

In the framework of a VKRS, a specific meaning is assigned to the *Closed-World Assumption*[5]: it connects the use of weak and strong negation in combination with partially and totally represented predicates. If the Closed-World Assumption holds for a predicate, its weak negation implies its strong negation, in other words, it is sufficient for an atomic sentence formed with such a predicate to be false if it is false by default.

It is shown that a *rule-based system* is a special kind of KRS which can also be viewed as an *information state* in the sense of Belnap [1977]. Since most AI systems can be classified as rule-based systems (e.g. deductive database and logic programming systems such as Datalog, LDL, or Prolog, and many expert systems such as MYCIN, or OPS5), or contain a rule-based system as a component (e.g. many terminological KR systems in the tradition of KL-ONE, such as BACK, and many expert sytems, such as BABYLON, KEE or NEXPERT), this emphasizes the fundamental character of the general concept of a KRS presented in chapter 2, and of Belnap's notion of an information state.

In chapter 6, I extend the framework of positive logic programs by adding strong negation and identify the resulting logic programming system as a fragment of Nelson's paraconsistent constructive logic. With the help of a newly introduced definition of a disjunctive normal form in constructive logic, I am able to characterize logic programs as the definite constituents of formulas of constructive logic. More precisely, it is shown that a logic program is equivalent to a Harrop formula with implication-free premises.

In chapter 7, a new system with two kinds of negation is developed. I extend the framework of normal logic programs by adding strong negation expressing explicit falsity. Viewed from another perspective, I extend the framework of constructive logic by adding weak negation expressing implicit falsity by default. Since there seems to be no general introduction rule for weak negation in rule-based systems, and further complications are created by the possibility of an infinite search space, the discussion is restricted to the case of function-free programs.

Finally, in chapter 8, various possibilities to apply and generalize the proposed concepts and methods are discussed. It is shown how disjunctive knowledge and conditional queries can be modeled within knowledge representation systems. And, as a generalization of the concept of an active database, a formal construction of an active KRS based on update actions is presented.

---

[5] Cf. [Reiter 1978].

1.2 Vivid Knowledge Representation and Logic Programming

The investigation of various extensions of positive logic programs shows that program clauses are fundamental logical expressions not because they are Horn clauses (this is rather a side effect of positive clauses) but because they represent rules with definite conclusions, or, equivalently, links of certain AND-OR-graphs.

Traditionally, work in logic programming has concentrated on the semantics of negation-as-failure. *Normal programs*[6] consist of clauses of the form

$$a_0 \leftarrow a_1, \ldots, a_m, -a_{m+1}, \ldots, -a_n$$

where $a_0, \ldots, a_n$ denote atoms and '$-$' stands for negation-as-failure.

Logic programming formalisms with two kinds of negation have independently been proposed in [Gelfond & Lifschitz 1990] and [Wagner 1990+1991a]. The idea of logic programming with two kinds of negation seems to become a new paradigm in knowledge representation and AI. Following [Gelfond & Lifschitz 1990], such programs are called *extended*. In their standard syntax they consist of clauses of the form

$$l_0 \leftarrow l_1, \ldots, l_m, -l_{m+1}, \ldots, -l_n$$

where $l_0, \ldots, l_n$ denote literals (formed by means of strong negation, $l = a | \sim a$).

There have been a great number of proposals on the semantics of normal logic programs, mostly in the form of fixpoint operators, or transformational interpretations (by reduction to positive programs), sometimes combined with 3-valued truth-functions, or in the form of translations to nonmonotonic formalisms such as default or autoepistemic logics. The most successful proposals seem to be the *stable model semantics* of [Gelfond & Lifschitz 1988], and the *wellfounded semantics* of [Van Gelder, Ross & Schlipf 1988]. It is an indication of the adequacy of a semantics for normal programs whether it can be carried over to extended programs in a more or less straightforward way. The benefit of the generalization to extended programs is a clearer semantical picture with respect to the notion of falsity. The semantics of falsity was rather obscured in many of the approaches to normal programs because there was no distinction between implicit falsity (by default) and directly established (constructible) falsity.

The principal case of logic programming is the one where programs do not contain negative loops,[7] and thus can be handled efficiently by an interpreter (i.e. an implemented inference procedure). These programs are called *weakly wellfounded*, or *locally stratified*. Although the semantics of weakly wellfounded

---

[6] Also called "general" in the literature.

[7] In fact, since standard Prolog systems implement a depth-first proof strategy, they cannot even handle positive loops.

programs is well-understood, and all approaches in the literature aggree on this, the vivid logic approach presented here is the only one which is firmly based on well-known logics by interpreting negations, conjunction and disjunction in partial logic, and interpreting the rule conditional '←' in constructive logic. This seems to be a clear advantage compared to the rather 'technical' character of many fixpoint and transformational semantics employing special (and very costly) techniques just to resolve negative loops which only rarely occur in practice.[8]

Another interesting approach is the interpretation of logic programs in a suitable modal logic. Obviously, since related to provability, the negation-as-failure operator has a modal flavor. Also the rule conditional '←', identified as first-degree constructive implication in [Pearce & Wagner 1989], can be interpreted in modal terms through the well-known translation of intuitionistic logic into modal S4 suggested by Gödel. This motivates the attempt to develop a nonmonotonic modal logic where in addition to a belief operator also negation-as-failure is treated as a modal operator (see e.g. [Lifschitz & Woo 1992]). But what is the gain of such a translational enterprise ? At the cost of introducing two modalities, negation and implication can be viewed as classical connectives again, and by subsuming default and autoepistemic logic the close connection of logic programming to both formalisms can be explained. However, from the point of view of pragmatics, where a semantics should be as simple and natural as possible, and help to find and validate effective derivation procedures, the introduction of modal operators seems to be a severe complication which might not be justified. At least for the principal case, i.e. for weakly wellfounded programs, one does not need a modal framework, and instead of a rather complex Kripke-style semantics a much simpler partial semantics is sufficient.

In any case, the point to be made is that a truly logical semantics of the prinicpal case of logic programming is of great importance and should be settled first, before the general case is considered. It is shown here that the semantics of weakly wellfounded extended logic programs can be given in terms of partial logic, and it is claimed that this partial semantics is the most natural one for such programs.

---

[8] In the general case, deciding whether a propositional program has a stable model is an NP-complete problem (see [Marek & Truszczynski 1991]), while for stratified programs there is a linear time algorithm computing the unique stable model (see [Niemelä & Rintanen 1992]). Therefore, it may eventually turn out that – similar to the case of the *occur check* – none of the proposed general semantics will be implemented in real logic programming systems since their overhead in treating the principal case of loop-free programs may, in practice, outweigh the gain in generality and loop-robustness.

Chapter 2

Vivid Knowledge Representation and Reasoning

2.1 Principles of Vividness

The notion of *vivid knowledge* has been introduced by Levesque in his "Computers and Thought" lecture at IJCAI-85. The paradigm for vivid knowledge is the kind of knowledge represented in a relational database. The contrary of it is disjunctive knowledge and other forms of indeterminate information which is also called 'incomplete'. While in [Levesque 1986] and [Etherington et al. 1989] the representation and processing of negative information is considered as one of the major problems for vivid reasoning,[1] I will show how to express explicit negative information in a vivid form avoiding the supposed 'incompleteness' of negation.

The idea of vivid knowledge representation and reasoning concerns two main aspects of AI technology: cognitive adequacy and computational feasibility. The goal is to develop an appropriate methodology for a knowledge representation and reasoning system (KRS) modeling certain forms of commonsense reasoning. The point of departure is the kind of knowledge representation achieved by relational and deductive databases (RDBs and DDBs[2]).

Related to the goal of cognitive adequacy is the methodological principle of

---

[1] For example, Etherington et al. [1989] note that "The major deficiency in the system as it stands is that it provides no mechanism for explicitly telling the VKB negative information."

[2] The concept of a DDB emphasizes the knowledge representation capabilities of logic programs. A DDB is a function-free logic program, that is, DDB theory deals with logic programming based on finite universes.

conceptual directness. It leads to the rejection of all kinds of indirect concepts and methods such as indirect proofs (by refutation of the contrary) or validity in all extensions (as e.g. in the ideally skeptical theory of defeasible inheritance). It also implies a principle of 'linguistic honesty': *What You Say Is What You Mean (WYSIWYM)*, stated by Sheperdson [1988]. In interpreting linguistic, resp. logical, expressions we are to take them as they are and not to impose any additional content on them just to accommodate them for some already well-established formalism, such as classical logic.

While in many (rather academic) approaches knowledge is represented as a set of arbitrary first-order formulas, a vivid approach seeks for inference systems based on appropriately restricted query and representation languages, thus taking into account the trade-off between expressiveness and efficiency. Both RDBs and DDBs can be considered as paradigms of real world knowledge representation. They implement a form of nonmonotonic reasoning caused by the use of negation-as-failure which deals with implicit negative information.[3] Although this negation operation clearly violates the semantics of classical standard logic, various attempts have been made to explain it in the framework of classical semantics, such as the completion semantics of Clark [1978]. As Sheperdson already noticed, besides other shortcomings the completion semantics blatantly violates the WYSIWYM principle.

The handling of negative information in databases and logic programs has turned out to be one of the biggest problems both practically and theoretically. Negation-as-failure allows for the querying and processing of implicit negative information only. It is not able to take care of explicit negative information. Although several authors have attempted to justify the restriction of the representation language to positive information,[4] this is questioned today. Notably Pearce and Wagner [1989] have argued that it would be more natural to treat negative information on a par with positive information, and therefore to allow for a means of direct reasoning to the explicit falsity of a query. A natural way to

---

[3] In the literature an RDB is sometimes considered to specify a classical first-order model. According to this view database tables are complete specifications of two-valued relations, and negation in database queries is classical negation, i.e. there is no nonmonotonicity. Since the dynamics of information change cannot be captured by classical two-valued models, and since it seems to be a gross overidealization to assume database tables to be always complete (see the discussion of inexact predicates in 4.3), a syntactic interpretation of RDBs as a set of ground atoms is certainly more adequate, and more natural.

[4] There seems to be no clear attitude towards the role of falsity, negative information and negation among AI researchers. For example, Przymusinski [1989, p.660] states that logic programs and deductive databases should be "free from excessive amounts of explicit negative information and as close to natural discourse as possible". And in [Rajasekar, Lobo & Minker 1989] it is maintained, that "explicit representation of negative information in logic programs is not feasible in many applications such as deductive databases and artificial intelligence". These statements are obviously questionable. First, logic programs and deductive databases do not usually permit one to represent explicit negative information at all. Secondly, explicit negative information clearly plays an important role in natural discourse, and hence, should be taken into consideration in vivid knowledge bases, deductive databases and logic programs.

accommodate this kind of negative reasoning in logic programs is the introduction of a strong negation operator expressing explicit (or, directly established) falsity, as proposed in [Pearce & Wagner 1989], and more elaborately in [Wagner 1991b]. A similar suggestion was made independently by Gelfond and Lifschitz [1990].

There are at least three different notions of falsity which are relevant in knowledge representation. Each of them provides the basis for defining a negation operator expressing the resp. kind of falsity in the object language.

Directly Established (Constructible) Falsity This is the notion of falsity modelled by Fitch, Nelson and Markov in their systems of constructive logic ([Fitch 1948+1952], [Nelson 1949], [Markov 1950]). The constructible falsity of a formula can only be established by a direct proof. The corresponding negation is called *strong* and is denoted by '$\sim$'. The strong negation of a sentence holds if the sentence is definitely (or explicitly) false.

Falsity as Inconsistency A formula is false if its addition to the knowledge base leads to inconsistency. The corresponding negation sign will be denoted by '$\neg$'. *Negation-as-inconsistency* occurs in several variants, for instance in Johannson's minimal logic and in Heyting's intuitionistic logic.[5]

Default-Implicit Falsity This is the notion of falsity underlying the negation-as-failure operator. A sentence is false by default if the assumption of its falsity is safe (i.e. definitely compatible with the available information). The corresponding negation is called *weak* and will be denoted by '$-$'.

An example illustrates the differences between the three kinds of negation:

Example 2.1   *Let*

$$KB = \begin{cases} q(c) \\ -q(d) \rightarrow \sim p(d) \\ r \rightarrow \sim q(c) \end{cases}$$

Since there is no possible argument for $q(d)$ on the basis of KB, $q(d)$ is implicitly false, and consequently, we obtain $-q(d)$. Hence, by a direct line of reasoning, using only detachment, $\sim p(d)$, i.e. the constructible falsity of $p(d)$, can be established. The falsity of r cannot be established directly but only by means of the indirect rule of (Negation as Inconsistency): the assumption of r leads to the contradiction $q(c), \sim q(c)$. Hence, we only get $\neg r$ but not $\sim r$. Notice that $q(d)$ is only false by default, $-q(d)$, but neither $\neg q(d)$ nor $\sim q(d)$ hold.

Logicians have invented various negations, however quite often only as a matter of mathematical exercise and not for practical purposes. Classical (or Boolean)

---

[5] A study of its use in logic programming can be found in [Gabbay & Sergot 1986].

negation seems to be the most idealized form of negation. It allows for such indirect – cognitively questionable and computationally costly – methods like *Reasoning by Cases* and *Reductio ad Absurdum*. Other negation concepts, like e.g. that of intuitionism, usually step away from it with respect to certain cognitive aspects where they claim to be more realistic. Interestingly, there is one negation concept which has emerged neither from mathematical exercise nor from philosophical considerations, but solely from computational practice in RDBs and normal logic programs: negation-as-failure. It will be viewed here as a specific form of the general concept of weak negation.

Both strong and weak negation are 'local' in the sense that their derivability check depends only on a certain restricted subset of the knowledge base whereas negation-as-inconsistency and classical negation require global search through the entire KB.

Remarkably, after more than twenty years of research, there is still no consensus on what is the proper meaning of negation-as-failure in general DDBs and logic programs. From practical experience, however, it is clear that any real world reasoning system should implement negation-as-failure in some way. The absence of it can be regarded as a severe shortcoming. This implies, of course, that the underlying logic must allow for nonmonotonicity. Closely connected to negation as failure is the Closed-World Assumption (CWA) which allows to infer negative queries from purely positive information. Since the CWA seems to be fundamental for database theory it must be accounted for as well in any vivid KRS.

Rather than taking some logical system known from the literature and then adapt it by certain modifications[6] the vivid reasoning research program starts from the scratch. By only taking those forms of reasoning for granted which have proved themselves in the framework of databases we get rid of all idealizations implicit in classical logic. As Levesque [1988] has put it,

> *The deviations from classical logic that will be necessary to ensure the tractability of reasoning stand in very close correspondence to the deviations from logic that we would have to make anyway to be psychologically realistic.*

Thus, the logic of vivid reasoning pursues a bottom-up theory formation process trying to avoid any theoretical overload. With regard to the great number of logical systems investigated since Frege, however, it would be quite unlikely not to find one which could serve as the base logic for our enterprise. Constructive logics with strong negation, as proposed by Fitch [1948+1952], Nelson [1949] and Markov [1950], seem to be best suited to the requirements of a vivid KRS. They have been developed as an alternative to classical logic which proved to be cognitively inadequate even for the business of doing mathematics. The main

---

[6] As it is done, for instance, with classical logic by cautiously adding some form of nonmonotonicity ending up with 'supraclassicality'.

concern of constructivists was to model cognitively relevant forms of negation and implication because exactly those operations have been extremely overidealized in classical logic.

Constructive logic can roughly be characterized as partial logic extended by adding intuitionistic implication. By dropping the *ex contradictione sequitur quodlibet* rule, a paraconsistent version of it is obtained which satisfies all requirements for vividness except that it does not possess a weak negation operator.

There are two outstanding theoretical contributions to the vivid reasoning research program I want to mention here.[7] First, Belnap [1976+1977] made significant suggestions towards a logic for vivid reasoning. He proposed to use a set of partial interpretations (he calls them 'set-ups') for the representation of extensional knowledge and to supplement it with a set of rules for the representation of intensional knowledge. He also gives a semantic account of logical operators in terms of both input and output of the proposed KRS, similar to the epistemic semantics of [Gärdenfors 1988]. Second, Levesque [1986+1988] introduced the terminology of 'vividness' and gave an outline of the principles and goals of vivid reasoning. I will discuss these foundational contributions further in the sequel.

2.2 Vivid Logic versus Standard Logic

I will call the logic of vivid knowledge representation and reasoning *vivid logic*. Notice, however, that this is not a fixed system but rather an evolving class of new logics based on Belnap's four-valued logic **B** and Nelson's paraconsistent constructive logic **N**.

The language of vivid logic consists of the logical operator symbols $\wedge, \vee, \sim, -$ and 1 standing for conjunction, disjunction, strong negation, weak negation and the verum, respectively; the predicate symbols $p, q, r, \ldots$; the constant symbols $c, d, \ldots$ and variables $x, y, \ldots$. Notice that there are no functional terms but only variables and constants. Further connectives, such as (possibly restricted) quantifiers, and epistemic modalities can be added to the language as an option.

An *atom* is an atomic formula, it is called *proper*, if it is not 1. *Literals* are either atoms or strongly negated atoms. *Extended literals* are either literals or weakly negated literals. I use $a, b, \ldots, l, k, \ldots, e, f, \ldots$, and $F, G, H, \ldots$ as metavariables for atoms, literals, extended literals and formulas, respectively.[8] A variable-free expression is called *ground*. The set of all proper ground atoms (resp. literals, resp. extended literals) of a given language is denoted by At (resp. Lit, resp. XLit). If not otherwise stated, a formula is assumed to be ground. If *op-list* is a set of logical operators, say *op-list* $\subseteq \{1, -, \sim, \wedge, \vee, \rightarrow\}$, then $L(op$-

---

[7]It seems likely that there are some more of which I am not aware, and so cannot do justice to them.

[8]I will frequently just say "literal" when I, precisely speaking, mean a proper ground literal.

list) denotes the respective set of well-formed formulas.

With each negation a complement operation for the resp. type of literal is associated: $\tilde{a} = \sim a$ and $\widetilde{\sim a} = a$; $\bar{l} = -l$ and $\overline{-l} = l$. These complements are also defined for sets of resp. literals $L \subseteq$ Lit and $E \subseteq$ XLit: $\tilde{L} = \{\tilde{l} : l \in L\}$, resp. $\overline{E} = \{\overline{e} : e \in E\}$.

Since the basic motivation for vivid logic is to provide an appropriate model for certain forms of practical reasoning one can hardly expect it to be as regular as standard logics. In standard logics concepts are often axiomatised according to idealised requirements originating in the domain of mathematics but unlikely to be suitable for commonsense reasoning. A standard consequence relation $\vdash$ is a relation between a set of premise formulas $X \subseteq L$ and a conclusion formula $F \in L$, for some fixed language L:

$$\vdash \,\subseteq\, 2^L \times L$$

Its axiomatization by Tarski requires it to satisfy the following conditions:

Reflexivity	$F \in X \Rightarrow X \vdash F$
Lemma Redundancy	$X \cup \{F\} \vdash G$ & $X \vdash F \Rightarrow X \vdash G$
Monotonicity	$X \vdash G \Rightarrow X \cup \{F\} \vdash G$

A vivid inference relation, however, is not uniform in general. Only certain formulas may make sense for representing vivid knowledge, that is, there will be a specific representation language $L_{\mathbf{Repr}}$, and a KB will be a finite collection of elements of $L_{\mathbf{Repr}}$, possibly constrained in some way determined by the set $L_{\mathbf{KB}}$ of all admissible KBs:

$$\text{KB} \in L_{\mathbf{KB}} \subseteq 2^{L_{\mathbf{Repr}}}$$

Likewise, since not every formula may be appropriate as a sensible query, the set of admissible queries is specified by $L_{\mathbf{Query}}$. A vivid inference relation, thus, is not based on a single universal language applying to premises as well as to queries (resp. consequences), but on two, usually different, languages:

$$\vdash \,\subseteq\, L_{\mathbf{KB}} \times L_{\mathbf{Query}}$$

Moreover, the axiomatization of standard consequence seems to be overidealised and not adequate for commonsense reasoning. This is no longer debated today in the case of the monotonicity postulate which requires that all previously obtainable answers remain valid whenever new information comes in. In many forms of commonsense reasoning, such as default reasoning, and in many computational systems, such as Prolog, monotonicity is violated, and one has to look for other principles replacing it, or one can just drop it and allow for unrestricted nonmonotonicity.[9]

---

[9] It is an open debate if, and in what form, nonmonotonicity should be restricted in a KRS. The currently most popular proposal is *Cautious Monotonicity* which will be called *Lemma Compatibility* below. Another, even stronger, restriction is *Rational Monotonicity*.

In a vivid KRS, also the reflexivity condition need not be valid in general. It might be desirable to exclude contradictions from it,

$$\{p, \sim p\} \not\vdash p,$$

in order to model defeasible reasoning.[10]

The basic scenario of a KRS consists of two operations: an inference, resp. answer, operation processing queries posed to the KB, and an update operation processing input formulas entered by users or by other (e.g. sensoric) information suppliers. While in standard logics an update is a simple addition of a formula $F \in L$ to the premise set $X \subseteq L$, i.e. $X \cup \{F\}$, a KRS restricts the admissible inputs to elements of a specific input language L_{Input}, and an update is performed by processing the input formula in an appropriate way in order to add its information content to the KB.

In general, a KB can consist of any kind of data structures capable of representing knowledge, e.g. a set, or multiset, or sequence, of (logical) expressions, or a directed graph, etc. For the sake of simplicity, I will assume that a KB $\in L_{\text{KB}}$ is a finite set of expressions from a representation language, i.e. $L_{\text{KB}} \subseteq 2^{L_{\text{Repr}}}$. There will always be an informationally 'empty' KB, denoted by 0, which is not necessarily equal to the empty set.

2.2.1 Inferences and Answers

An inference relation of a KRS is an effectively computable relation between KBs and closed queries $Q \in L_{\text{Query}}$, $\text{KB} \vdash Q$, establishing that Q holds, or is derivable, on the basis of KB. An answer operation **Ans** takes a knowledge base and a query formula, and provides an answer in L_{Answer}:

$$\textbf{Ans} : L_{\text{KB}} \times L_{\text{Query}} \longrightarrow L_{\text{Answ}}$$

An answer can be a relation, collecting all valid substitutions for Q, in the case of an open query, or it can be an element of a set of answer values containing at least **yes** and **no**, in the case of a closed query:

$$\textbf{Ans}(\text{KB}, Q) = \textbf{yes} \stackrel{def}{\Longleftrightarrow} \text{KB} \vdash Q,$$

if Q is closed,[11] and otherwise,

$$\sigma \in \textbf{Ans}(\text{KB}, Q) \stackrel{def}{\Longleftrightarrow} \text{KB} \vdash Q\sigma$$

---

[10] I differ here from Makinson [1989] who argues that procedures that do not satisfy this condition "are best thought of not as processes of 'inference' but rather of theory or knowledge change."

[11] More precisely, if the query language allows for negation, $\textbf{Ans}(\text{KB}, Q) = \textbf{yes} \stackrel{def}{\Longleftrightarrow} \text{KB} \vdash Q$ & $\text{KB} \not\vdash \sim Q$.

where $Q\sigma$ denotes the ground substitution instance of Q under the substitution σ.

Example 2.2   *For* KB $\subseteq$ At *and* $L_{\text{Query}} = L(-, \wedge, \vee)$ *define the inference relation* $\vdash_{RDB}$ *as follows:*

$$
\begin{array}{lll}
\text{KB} \vdash_{RDB} a & \text{if} & a \in \text{KB} \\
\text{KB} \vdash_{RDB} -a & \text{if} & a \notin \text{KB} \\
\text{KB} \vdash_{RDB} F \wedge G & \text{if} & \text{KB} \vdash_{RDB} F \ \& \ \text{KB} \vdash_{RDB} G \\
\text{KB} \vdash_{RDB} -(F \wedge G) & \text{if} & \text{KB} \vdash_{RDB} -F \ \text{or} \ \text{KB} \vdash_{RDB} -G \\
\text{KB} \vdash_{RDB} F \vee G & \text{if} & \text{KB} \vdash_{RDB} F \ \text{or} \ \text{KB} \vdash_{RDB} G \\
\text{KB} \vdash_{RDB} -(F \vee G) & \text{if} & \text{KB} \vdash_{RDB} -F \ \& \ \text{KB} \vdash_{RDB} -G \\
\text{KB} \vdash_{RDB} --F & \text{if} & \text{KB} \vdash_{RDB} F
\end{array}
$$

Let $\text{KB}_1 = \{p(a), q(a, b), q(b, c)\}$. *Then, for instance,* $\text{KB}_1 \vdash_{RDB} -(q(a, a) \wedge -p(a))$, *and*

$$\textbf{Ans}(\text{KB}_1, q(x, y) \wedge -p(x)) = \{\langle b, c \rangle\}$$

An inference operation C can be associated with $\vdash$ in the usual way:

$$C(\text{KB}) = \{F \in L_{\text{Query}} : \text{KB} \vdash F\}$$

2.2.2   Updates

An update operation **Upd** takes a KB and an input formula $F \in L_{\text{Input}}$, and provides an appropriately updated KB:

$$\textbf{Upd} : L_{\text{KB}} \times L_{\text{Input}} \to L_{\text{KB}}$$

An update may be simply an addition of a formula $F \in L_{\text{Input}} \cap L_{\text{Repr}}$[12] to the KB,

$$\textbf{Upd}(\text{KB}, F) = \text{KB} \cup \{F\}$$

in which case the symbol $\cup$ may be used instead of **Upd**. In general, it will be necessary to process the input formula in some way, especially if it is not an L_{Repr} expression. For instance, an RDB can be updated by a conjunction of atoms by simply adding all atoms to it.

Example 2.3   *Let* KB $\subseteq$ At, *then*

$$\textbf{Upd}(\text{KB}, \bigwedge_{i=1}^{n} a_i) = \text{KB} \cup \{a_1, \ldots, a_n\}$$

---

[12]Notice that L_{Input} and L_{Repr} may be totally distinct. One may expect that an input formula is a kind of logical expression whereas a KB can be any kind of data structure.

Updates may be constrained by *integrity constraints*. Following [Reiter 1990], we define a set of integrity constraints specified for a certain KB as any set of closed query formulas, $IC \subseteq L_{\text{Query}}$, such that an input $F \in L_{\text{Input}}$ is considered *illegal* whenever $\mathbf{Upd}(\text{KB}, F) \not\vdash G$ for some constraint $G \in IC$. Update attempts violating a constraint have to be rolled back by the system.

2.2.3 Knowledge Representation Systems

Thus, in a narrower sense, a KRS is a triple:[13]

$$\langle L_{\text{KB}}, \vdash, L_{\text{Query}} \rangle$$

In general, however, a KRS is a quintuple:

$$\langle \mathbf{Ans}, L_{\text{KB}}, L_{\text{Query}}, \mathbf{Upd}, L_{\text{Input}} \rangle$$

which will also be formulated as

$$\langle L_{\text{KB}}, \vdash, L_{\text{Query}}, \mathbf{Upd}, L_{\text{Input}} \rangle$$

Example 2.4   *The KRS*

$$\langle 2^{\text{At}}, \vdash_{RDB}, L(-, \wedge, \vee), \cup, \text{At} \rangle$$

is denoted by $\mathbf{V}_{\text{RDB}}$ *since it can be viewed as a formal description of the logical structure of relational databases.*

The formulation of a KRS in terms of query and input processing was already implicitly present in Belnap's view of a KRS [Belnap 1977]. In [Levesque 1984a] it was proposed as a 'functional approach to knowledge representation'.[14]

2.2.4 Examples

Prominent instances of this KRS scenario are:

Classical Logic Here, the KB consists of a (possibly infinite) set of arbitrary first order sentences, and is usually called a 'theory', Q is an arbitrary (possibly open) formula, and an answer is either **yes** or **no**, or a verifying substitution for Q. The inference relation is determined by two-valued

---

[13] In [Herre & Pearce 1992] a similar notion of a 'calculus as a triple' is introduced with the same motivation: in order to single out certain restrictions of the representation and query language.

[14] However, Levesque is not so much concerned with the investigation of the formal properties of KRSs in general, but rather with the modelling of the concept of meta-knowledge by means of an epistemic modality on the basis of classical logic. See also the remarks on Levesque's K operator in 2.5.1.

truth-functions (leaving no room for the distinction between different types of negation). It satisfies monotonicity. Updates are simple additions of formulas. Thus, the representation language is equal to the query and input language, $L_{\text{Repr}} = L_{\text{Query}} = L_{\text{Input}}$.

Relational Databases The representation language of RDBs admits only of positive facts, i.e. ground atoms, represented as tuples of certain relations. The query language, in practice, is SQL. It could as well be described as the language of $\land, \lor, \textbf{not}, \exists, \forall$ (where **not** denotes negation-as-failure). Answers are relations, also called views. The answering mechanism consists of table lookups and relational algebra operations. It behaves nonmonotonically since **not** is interpreted as set difference within relational algebra. Updates are mainly insertions and deletions of tuples from relations.

Deductive Databases The representation language admits only of positive facts and rules of the form $a \leftarrow l_1, \ldots, l_n$, where a is an atom, and l_i are literals (formed with negation-as-failure). A query is a conjunction of literals, and an answer is either **yes** or **no**, or a verifying substitution. The inference operation of DDBs is traditionally described as some form of linear resolution. More recently, bottom-up procedures in combination with certain pre-compilation techniques flattening the hierarchical structure of nested rules, such as the 'Magic Set' or 'Alexander' method, seem to be preferred because of greater efficiency (see e.g. [Bry 1990]). Declaratively, the semantics of DDBs is either based on the unique 'natural' model defined inductively in the 'stratified' case, or on the wellfounded (cf. [Van Gelder, Ross & Schlipf 1988]), or stable (cf. [Gelfond & Lifschitz 1988]) semantics in the general case. In any case, it is nonmonotonic.

Terminological Knowledge Representation Systems in the tradition of KL-ONE maintain a KB consisting of two separate parts: the *terminological* KB ('TBox') stores terms and their definitions, while the *assertional* KB ('ABox') stores sentences constructed using these terms. For example, the BACK system[15] accepts as terminological input concept declarations and definitions, like

FATHER := PARENT ⊓ MALE,

and as assertional input object descriptions, like

Tom :: FATHER.

A specific update operation[16] for terminological input performs certain normalization steps in order to transform definitions into a canonical form

---

[15] See [Quantz & Kindermann 1990].
[16] Called *DEFINE* in [Levesque 1984a].

and places new concepts within the taxonomic hierarchy maintained in the TBox according to their subsumption relationships with already present concepts (classification). Queries are either term subsumption formulas, like

FATHER? < PARENT,   *("Is a father a parent ?")*

or object recognition formulas, like

Tom? : MALE,   *("Is Tom male ?").*

Subsumption formulas are inferred by means of a 'normalize-compare' algorithm: in the first step terms are transformed into normal forms, and in the second step these normal forms are compared. The logic involved in this inference procedure seems to be subclassical: negation is only allowed in a very restricted form.

Notice that classical logic is a special case of a KRS, and not the other way around. RDBs and DDBs are other instances of KRSs. However, they are not special cases of classical logic.

Refering to the three main parameters in vivid logic, namely L_{KB}, L_{Query} and $\vdash$, the essential question is:

What kind of information can be extracted from what forms of sentences in a computationally dependable way ? (Levesque, 1986)

2.3   Different Types of Inference

I already mentioned that the axiomatisation of logical consequence due to Tarski is not adequate for commonsense reasoning. Both the condition of reflexivity and monotonicity will have to be relaxed.

First, reflexivity should be restricted to consistent formulas.[17] This takes into consideration that in theories of defeasible reasoning contradictory facts invalidate each other, thus causing both irreflexivity and nonmonotonicity.

Second, monotonicity should either be dropped completely, or replaced by a weaker condition such as *Cautious Monotonicity*, due to Gabbay [1985], which is called *Lemma Compatibility* here, since it guarantees that the addition of a lemma to the knowledge base does no harm to further inferences. Notice that there are nonmonotonic formalisms, such as Reiter's default logic, or *disjunctive fact bases* (see 4.6), which do not satisfy Lemma Compatibility. Therefore, it seems to be too restrictive to make Lemma Compatibility a strict requirement.

---

[17] Although most logicians will find this restriction strange, I am not in bad company: both Bolzano [1837] (see [van Benthem 1984]) and Wittgenstein [1956] have already proposed it.

Rather, it should be considered as a useful property enjoyed by some systems but violated in others.

Third, additional conditions for a vivid inference relation will have to be formulated, notably *Constructivity* and *Non-Explosiveness*.

Let $F \in L_{\text{Input}} \cap L_{\text{Query}}$ and $G \in L_{\text{Query}}$.

Restricted Reflexivity   If F is consistent in KB,[18] then

$$\mathbf{Upd}(\text{KB}, F) \vdash F$$

Lemma Redundancy (alias *Cut*, or *Transitivity*)

$$\text{KB} \vdash F \ \& \ \mathbf{Upd}(\text{KB}, F) \vdash G \ \Rightarrow \ \text{KB} \vdash G$$

Lemma Compatibility (alias *Cautious Monotonicity*)

$$\text{KB} \vdash F \ \& \ \text{KB} \vdash G \ \Rightarrow \ \mathbf{Upd}(\text{KB}, F) \vdash G$$

Lemma Redundancy and Compatibility can be combined in the following condition of

Cumulativity

$$\text{KB} \vdash F \ \Rightarrow \ C(\mathbf{Upd}(\text{KB}, F)) = C(\text{KB})$$

The following conditions of Rational Monotonicity and Monotonicity seem to be too strong for most KRSs. In special cases, however, they may hold.

Rational Monotonicity

$$\text{KB} \nvdash \sim F \ \& \ \text{KB} \vdash G \ \Rightarrow \ \mathbf{Upd}(\text{KB}, F) \vdash G$$

Monotonicity

$$C(\text{KB}) \subseteq C(\mathbf{Upd}(\text{KB}, F))$$

Clearly, Monotonicity implies Lemma Compatibilty.

2.3.1   Constructivity

The notions of *constructible truth* and *constructible falsity* stem from intuitionistic, resp. constructive, logic where they are called *disjunction property*, resp. *constructible falsity*.[19]   While the disjunction property expresses the constructivity (resp. decomposability) of truth for disjunctions: a disjunction can only

---

[18] According to an appropriate notion of consistency associated with each KRS.
[19] See e.g. [Rautenberg 1979].

be true if one of its disjuncts is, the property of constructible falsity expresses the constructivity of falsity for conjunctions: a conjunction can only be false if one of its conjuncts is. In traditional logics these properties refer to tautological truth and falsity. In vivid logic we are not so much concerned with tautologies but rather with the validity of inferences. The following definitions, therefore, are appropriate generalizations of the original ones.

Let $L \subseteq$ Lit be an arbitrary set of ground literals (with respect to the principal negation $\sim$), and $\lceil L \rceil := \mathbf{Upd}(0, \bigwedge L)$. Then $\vdash$ is called *constructive* if it satisfies (i) constructible truth, and (ii) constructible falsity, i.e. both of the following conditions,

(i) $\lceil L \rceil \vdash F \vee G \;\Rightarrow\; \lceil L \rceil \vdash F \;$ or $\; \lceil L \rceil \vdash G$

(ii) $\lceil L \rceil \vdash \sim (F \wedge G) \;\Rightarrow\; \lceil L \rceil \vdash \sim F \;$ or $\; \lceil L \rceil \vdash \sim G$

Obviously, the first condition (of constructible truth) excludes the possibility of certain disjunctive tautologies such as the *tertium non datur*, $F \vee \sim F$, whereas its negative counterpart excludes, for instance, the *principle of contradiction*, $\sim (F \wedge \sim F)$.

Observation 2.1

(i) *Classical logic is not constructive since, for instance, $\{q\} \vdash p \vee \sim p$, or, $\{p\} \vdash (q \rightarrow r) \vee (r \rightarrow q)$.*

(ii) *Johannson's minimal and Heyting's intuitionistic logic are not constructive since, for instance, $\{q\} \vdash \sim (p \wedge \sim p)$.*

Constructivity is a necessary requirement for an inference relation to render formulas decomposable in the sense, that it suffices in order to establish the validity of a formula, to check the validity of certain subformulas. 'Complete' knowledge, according to Levesque [1986], is decomposable.

2.3.2   Non-Explosiveness

Every KRS needs to implement certain principles in order to deal with contradictory information in some way. In traditional logics it is prohibited to do business as usual in the presence of contradictions. The system is forced to break down instead, or, to make appropriate revisions. A pragmatic approach, however, simply recognizes that many of our beliefs, judgements, scientific theories, legal codes and so on actually turn out to be inconsistent. And it should be rather an option, and not an immediate obligation, to make respective revisions. It seems to be an unnatural overreaction to abandon a KB once it is discovered to be inconsistent. Rather one should accommodate it by means of a logic which continues to function plausibly under the burden of inconsistency.

Paraconsistency A KRS is called *paraconsistent* if the principle (ECSQ)[20] is
not valid, i.e. if $\{p, \sim p\} \not\vdash q$.

The rejection of inferences based on ECSQ is also motivated by the requirement of shared propositional variables (i.e. commonality of content) between
premises and conclusions which is essential in relevance logics. This requirement
seems to be particularly natural in the context of vivid reasoning. It is the basis
for the following definition (due to Urbas [1990]) which excludes the possibility
of non-tautological inferences from certain KBs solely based on the form of a
query and not on its content.

Non-Explosiveness Let F be any non-tautology (i.e. KB $\not\vdash F$ for some KB)
without weak negation, then $\vdash$ is called *non-explosive* if for all KBs there
is a variant F' of F (obtained by uniform substitution of propositional
constituents) such that KB $\not\vdash F'$.

Observation 2.2 *Johannson's minimal logic is explosive, since it supports
inferences of the sort* $\{p, \sim p\} \vdash \sim q$, *i.e. from a contradiction follows every
negation.*

Non-explosiveness generalizes paraconsistency in the direction of relevance logic.
However, the postulate of relevance logic that every inference has to be 'relevant' seems to be too strong a requirement for vivid reasoning. It would exclude
inferences based on *weakening*, like $\{p\} \vdash q \rightarrow p$. Urbas argues that in order to
retain certain desirable properties of negation, such as contraposition, one has
to make the second step, abandon weakening, after having made the first, requiring non-explosiveness, and thus one really ends up with relevance logic when
the starting motivation was only paraconsistency. Yet in the context of vivid
reasoning where many popular regularity properties of negation, like for instance
contraposition, are questioned, this argument is not convincing. Therefore, to
require inferences (and any possible implication to be introduced) to be relevant
is too restrictive for our purposes.

   Table 2.1 is a checklist for the degree of vividness of some well-known logical
systems which are denoted by the name of their proponents ('Heyting' stands for
intuitionistic logic, 'Johannson' for his minimal logic [Johannson 1937], 'Kleene'
stands for his strong 3-valued logic [Kleene 1952], 'Belnap' for his useful 4-valued
logic [Belnap 1977], and 'Nelson' for his paraconsistent constructive logic [Nelson
1949]).

2.4   Different Types of Knowledge Bases

While RDBs only allow for the representation of positive facts (i.e. ground
atoms), DDBs extend this rather restricted framework by also allowing for

---

[20] Ex contradictione sequitur quodlibet.

	constr. truth	constr. falsity	paracons.	non-explosive
Classical				
Heyting	√			
Johannson	√		√	
Kleene	√	√		
Belnap	√	√	√	√
Nelson	√	√	√	√

Table 2.1: Vividness criteria for some well-known logics.

the representation of positive conditional information in the form of rules with atomic conclusions. By employing negation-as-failure in the premise these rules are capable of also taking into consideration implicit negative information.

In the sequel, the framework of RDBs and DDBs will be extended to accommodate explicit negative information in a straightforward way by the introduction of *strong negation*. Thus, it becomes possible to represent

1. definite extensional information, by a set of (positive and negative) facts, i.e. a set of ground literals, also called the *extensional database*,

2. conditional information, by means of rules with possibly negative conclusions and referring to all other kinds of information (by means of weak negation, strong negation, conjunction and disjunction) in the premise, also called the *intensional database*.

An arbitrary set of truth-functional sentences cannot be represented by a single set of ground literals, i.e. a partial interpretation, in general. The reason for this is that the representation of disjunctive information requires a set of partial interpretations representing all disjunctive alternatives which are implicit in the given set of sentences. Since vivid knowledge representation concentrates on definite information, one cannot expect that a vivid knowledge base is capable of accepting arbitrary input. In particular, disjunctive input formulas may have to be excluded.

It will be useful to compare knowledge bases in terms of their information content, that is, to have an informational (pre)ordering between KBs such that

$$KB_1 \le KB_2 \quad \text{if } KB_2 \text{ contains at least as much information as } KB_1.$$

The informational ordering '$\le$' should be defined in terms of the structural components of knowledge bases and not in terms of higher-level notions (like derivability).

The informationally empty KB will be denoted by 0. By definition,

$$0 \le KB \quad \text{for all KBs.}$$

Definition 2.1 *A KB is called* input-constructible *if it can be represented by an appropriate input formula, that is, if*

$$\exists F \in L_{\text{Input}} : \text{KB} = \text{Upd}(0, F)$$

A KB is called input-destructible *if it can be contracted by an appropriate input formula, that is, if*

$$\exists F \in L_{\text{Input}} : \text{Upd}(\text{KB}, F) = 0$$

2.4.1 Unique Representation

A KRS enjoys the property of *Unique Representation* if the information contained in a KB is uniquely represented, that is, if

$$C(\text{KB}_1) = C(\text{KB}_2) \implies \text{KB}_1 = \text{KB}_2$$

Obviously, a standard logical system where a KB is a set of formulas does not enjoy the Unique Representation property. On the other hand, a relational database is a unique representation.

It seems to be desirable for a KRS that KBs are unique (or nearly unique) representations because this means that input formulas have to be 'compiled' into their (unique) standard representations thereby facilitating subsequent inferences. Unique representation is closely connected to the normalization of input formulas by transforming them into some (e.g. disjunctive) normal form. Reasoning on the basis of compiled (i.e. normalized) information seems to be computationally more effective. This is indicated, for instance, by the fact that the time complexity of the problem of entailment between fomulas in Belnap's 4-valued propositional logic is reduced from co-NP-complete to polynomial if both the premise and the conclusion are normalized (see [Patel-Schneider 1990]).

2.4.2 Definite Information

A KB is called *definite* with respect to a constructive inference relation ⊢, if a disjunction cannot be inferred unless one of its disjuncts can,

$$\text{KB} \vdash F \vee G \implies \text{KB} \vdash F \text{ or } \text{KB} \vdash G.$$

Notice that if the DeMorgan laws hold (which will be the case in all systems under consideration), this implies that a negated conjunction can only be inferred if the negation of one of its conjuncts can,

$$\text{KB} \vdash \sim(F \wedge G) \implies \text{KB} \vdash \sim F \text{ or } \text{KB} \vdash \sim G.$$

One could ask now:

Question 1   *Given an inference relation and a set of associated logical operators, say* op-list, *is there any specific form of a KB, i.e. restrictions on* $L_{KB} \subseteq 2^{L(op-list)}$, *guaranteeing definiteness ?*

For instance, if the KB consists only of literals, or conjunctions of literals, this is obviously the case. But what is the largest (least restrictive) language for KBs still guaranteeing definiteness ? Well-known is the class of Harrop formulas in intuitionistic logic which can also be defined in constructive logic. A set of Harrop formulas in constructive logic forms a definite knowledge base (see 3.3.3).

Clearly, disjunctive information is not definite. But it does not suffice to ban disjunctions from KBs in order to ensure definiteness. Indefinite information may also be represented by certain combinations of seemingly definite operators, like $\sim(F \wedge G)$, or in connection with other operators, like quantifiers, or negation-as-failure.

2.4.3   Negation-as-Failure and Indefinite Information

Negation-as-failure was born in the area of database reasoning. As discussed above, relational databases represent definite information (unless they admit null values). The meaning of negation-as-failure in the context of RDBs without null values is clear (yet only within definite query formulas, see below). A fact, say p, may be explicitly represented, $p \in RDB$, and consequently $RDB \vdash p$. If a fact, say q, is not represented in RDB, $q \notin RDB$, then $RDB \nvdash q$, and consequently $RDB \vdash -q$ where '$-$' stands for negation-as-failure. In both cases the conclusions inferred from the given information are compatible with it.

Difficulties with the notion of negation-as-failure arise, for instance, if the KB contains disjunctive information, as in the case of KB $= \{p \vee q\}$. Since neither p nor q is provable from KB, we obtain $\{p \vee q, -p, -q\}$, by a 'naive' negation-as-failure approach. Any reasonable interpretation of this set would require that $-p \vee -q$ does not imply $-(p \wedge q)$, for otherwise we would obtain $-p \wedge -q$ along with $-(-p \wedge -q)$ (i.e. failure and non-failure at the same time). But this violation of one of the DeMorgan laws also violates our intuitions about provability (and the related notion of failure to prove): if p or q are not provable their conjunction should also not be provable.

When negation-as-failure was implemented in Prolog by Colmerauer, its use was intended for definite databases only. Because Prolog does not allow one to represent disjunctive information, indefiniteness cannot be created by disjunction. However, a certain interaction between negation-as-failure and the rule operator can make a Prolog program indefinite. Take, for example, $\Pi = \{p \leftarrow -q, q \leftarrow -p\}$. Neither p nor q are derivable, but if we therefore add their resp. negations, we obtain an absurd information state, $\{-p, -q, p, q\}$. On the other hand, one could argue that $\Pi \vdash p \vee q$, i.e. Π is indefinite (for instance, by appealing to the principle *reasoning by cases*, or to minimal models).

The problem of identifying appropriate classes of logic programs for which negation-as-failure actually works and a logical explanation of this can be given seems to be related to the question as to which syntactic form of a program ensures its definiteness. For instance, the class of *locally stratified* programs is definite.[21] But it seems that local stratification will not be the last word in the ongoing enterprise of generalizing negation-as-failure.[22]

2.5   Different Types of Queries

Usually one would expect the query language to be at least as general as the representation language because it should be possible to ask everything that can be represented. On the other hand, even if the KB does not admit e.g. disjunctive information, disjunctive queries might be asked. Thus, it seems reasonable to have a query language which is more expressive than the representation language, as for instance in RDBs.

If derivability is defined for ground literals and weakly negated ground literals, and restricted to definite KBs, it can be extended in a natural way to arbitrary ground formulas $F \in L(1, -, \sim, \wedge, \vee)$ as described below in 2.10. Such query formulas are called *extensional* since they are formed with the help of truth-functional operators.

In general, it will not suffice in order to infer $-F$ that F fails (this will only be the case for definite KBs). However, the following restriction characterizes weak negation:

$$(\text{Inherent Consistency}) \quad \text{KB} \vdash -F \;\Rightarrow\; \text{KB} \not\vdash F$$

Notice that this does not hold for classical negation. In fact, Inherent Consistency is violated by every negation satisfying (ECSQ). Also strong negation, in general, does not satisfy Inherent Consistency. Yet, it seems desirable that the following *coherence*[23] property would hold:

$$(\text{Coherence}) \quad \text{KB} \vdash \sim F \;\Rightarrow\; \text{KB} \vdash -F \;\Rightarrow\; \text{KB} \not\vdash F$$

Positive answers are not necessarily preserved under growth of information. Queries, for which this is the case, are called *persistent*.

Definition 2.2 (Persistent Query)   *A closed, resp. open, query formula F is called* persistent *if* $\text{KB}_1 \vdash F \;\Rightarrow\; \text{KB}_2 \vdash F$, *respectively* $\text{Ans}(\text{KB}_1, F) \subseteq \text{Ans}(\text{KB}_2, F)$, *whenever* $\text{KB}_1 \leq \text{KB}_2$. *A KRS and its inference relation is called* persistent *if every query $F \in L_{\text{Query}}$ is persistent.*

---

[21] I will call these programs *weakly wellfounded* according to the definition in 7.5.

[22] E.g. in [Przymusinska & Przymusinski 1990] a further generalization of local stratification, called *weak stratification*, is presented.

[23] The name is adopted from [Pereira & Alferes 1992].

2.5.1   Intensional Queries

It may be desirable to have certain intensional operators, such as implication, or epistemic modalities, in the query language. Because their semantics depends on the semantics of the extensional operators problems with the latter have to be settled first.

A straightforward definition of implication in query formulas would be:

$$\mathbf{Ans}(\mathrm{KB}, F \rightarrow G) := \mathbf{Ans}(\mathbf{Upd}(\mathrm{KB}, F), G)$$

If the input language does not allow for arbitrary formulas (which would normally be the case) the premise formulas of such implications would have to be restricted in some way (e.g. they would have to be disjunction-free).

Levesque [1984b] proposes to use an epistemic modality K for expressing queries about definite and indefinite existence. Consider the following example,

$$\mathrm{KB} = \{p(a), q(b) \vee q(c)\}$$

For any standard inference relation $\vdash$, we would obtain $\mathrm{KB} \vdash \exists x\, p(x)$ as well as $\mathrm{KB} \vdash \exists x\, q(x)$. Although there is a clear difference between both queries (the former but not the latter can be answered definitely), this cannot be distinguished by means of a simple existential query. However, using Levesque's K modality, we obtain

$$\mathrm{KB} \vdash \exists x\, (p(x) \wedge K p(x))$$

and on the other hand,

$$\mathrm{KB} \vdash \exists x\, (q(x) \wedge \neg K q(x))$$

expressing the difference that in the first case there is an object for which KB knows that p, whereas in the second case there is an object for which q holds, but the KB does not know this object.

As is shown in [Patel-Schneider 1990] the first-order version of Belnap's four-valued logic can be made decidable by restricting the semantics of the existential quantifier to its definite interpretation (that is, giving up its reading as possibly infinite disjunction). Essentially, Patel-Schneider's existential quantifier $\exists'$ can be rewritten as

$$\exists' x\, p(x) \Longleftrightarrow \exists x\, (p(x) \wedge K p(x))$$

2.5.2   Non-ground Queries and the Problem of Definiteness

Non-ground queries in RDBs are relational calculus expressions, i.e. sentences or open formulas of a first order logic with negation-as-failure. It is a non-trivial problem in this context to determine which relational calculus expressions can be answered sensibly. An answer is called *sensible* if its values are limited to constants that occur in the query or in the KB. For instance, $-p(x)$ cannot be answered sensibly because it holds for arbitrary x's that are not in the KB. Also, $p(x) \vee q(x, y)$ is not sensible, it holds for arbitrary y values when $p(x)$ is true. On the other hand, $\exists y(p(x) \vee q(x, y))$ is sensible; it can be translated into the relational algebra expression $p \cup \pi_x(q)$ where π means projection, i.e. the answer is a set of 1-tuples, namely the collection of all objects for which p holds together with all objects occuring in the first column of q.

A sensible query is definite in the following sense. Let $\mathcal{I}$ be a standard (first-order) interpretation, and let $\mathcal{I}^*$ be the interpretation obtained from $\mathcal{I}$ by extending its universe by the addition of a new value $c_* \notin \mathrm{dom}\mathcal{I}$, such that $\mathcal{I}$ and $\mathcal{I}^*$ agree on all predicates in the KB. Then a query formula F is called *definite* if

$$\mathcal{I} \models F\sigma \quad \text{iff} \quad \mathcal{I}^* \models F\sigma,$$

i.e. F gets the same answers in both interpretations.

The class of definite formulas was shown to be not recursive.[24] So the problem is to define appropriate subclasses of it which can be handled efficiently by a query processor. Van Gelder and Topor [1991] give an inductive definition of the class of *evaluable formulas* which are definite and can be transformed into relational algebra expressions. They argue that the evaluable formulas form the largest practical subclass of the definite formulas.

2.5.3   Existential Queries and the Generation of Answers

Although the further considerations in this book are mostly restricted to the case of ground queries, I want to discuss briefly the topic of existential query formulas.

In general, the derivability of an existential formula, $\mathrm{KB} \vdash \exists x F(x)$, does not guarantee that an appropriate answer t (either definite or indefinite), such that $\mathrm{KB} \vdash F(t)$, can be generated by the given inference operation. This *Existential Property* is also the main issue of constructive mathematics where – unlike in the field of database theory – one usually deals with infinite KBs (i.e. recursively enumerable theories).

In [Herre & Pearce 1992] the notions of *constructive correctness* and *constructive completeness* capturing the Existential Property are introduced. The following definitions are adapted versions of these notions. Let the logic **L** be given by its consequence relation $\models_L$.

Definition 2.3   *A KRS is called* constructively correct *with respect to* **L**, *if*

---
[24] See [DiPaola 1969] and [Vardi 1981].

KB ⊢ ∃x F(x) *implies that there is a ground term (tuple) t such that* KB ⊢ F(t), *and t can be generated by the inference procedure* ⊢, *and* KB ⊨_L F(t).

Definition 2.4   *Let t be a ground term (tuple). A KRS is called* constructively complete *with respect to* **L**, *if* KB ⊨_L F(t) *implies that* KB ⊢ F(t), *and t can be generated by the inference procedure* ⊢.

For instance, if KBs consist of Horn formulas, L_{Query} is the set of existentially closed formulas whose matrix is a conjunction of atoms, and ⊢ is determined by SLD-resolution, then $\langle L_{\text{KB}}, \vdash, L_{\text{Query}}, \rangle$ is constructively correct and complete with respect to classical logic.

2.6   Different Types of Updates

The semantics of answering and updating depend on each other as manifested by conditions on their dynamic behaviour (see, for instance, the conditions of Restricted Reflexivity, Lemma Redundancy and Lemma Compatibility). The dichotomy of answering and updating has an obvious resemblance with the dichotomy of introduction and elimination in Gentzen sequent and natural deduction calculi. Restricted Reflexivity expresses a kind of adequacy property for the update operation: the processing of an input formula should respect its meaning as an inferrable query (notice the resemblance with the fact that elimination rules have to respect the corresponding introduction rules).

Definition 2.5 (Ampliative Input)   *An input formula F is called (i)* ampliative[25] *if* KB ≤ **Upd**(KB, F), *and (ii)* reductive *if* KB ≥ **Upd**(KB, F). *A KRS and its update operation* **Upd** *is called* ampliative *if all inputs* $F \in L_{\text{Input}}$ *are ampliative.*

If a KRS is monotonic, its update operation is ampliative.

Observation 2.3   *Ampliative updating is the formal counterpart of persistent answering.*

The structural rules of *Contraction* and *Permutation* are fundamental in Gentzen-style sequent systems where they essentially determine the logical structure of the premise of a sequent.[26] In a KRS, they are properties of the update operation with respect to the internal structure of the KB.

Contraction

$$\mathbf{Upd}(\mathbf{Upd}(KB, F), F) = \mathbf{Upd}(KB, F)$$

---

[25] The name is adopted from [Belnap 1977].

[26] See, e.g., [Wansing 1992].

Permutation

$$\mathbf{Upd}(\mathbf{Upd}(KB, F), G) = \mathbf{Upd}(\mathbf{Upd}(KB, G), F)$$

Observation 2.4   *The following condition of* Conjunction Composition,

$$\mathbf{Upd}(KB, F \wedge G) = \mathbf{Upd}(\mathbf{Upd}(KB, F), G)$$

can be used as a natural definition of conjunction processing. That is, with its help L_{Input} can always be closed under conjunction. Because standard conjunction is commutative, Conjunction Composition can only be used as a definition of conjunctive input processing if $\mathbf{Upd}$ satisfies Permutation, that is, if time does not matter for updates. In chapter 4 a simple system, V_{o}, will be presented where Permutation is violated, and therefore Conjunction Composition does not hold.

Observation 2.5   *The expressiveness of L_{KB} does not matter. More important is the expressiveness of the input and the query language as the following construction of the entailment relation associated with a KRS shows. For $F \in L_{\text{Input}}$ and $G \in L_{\text{Query}}$ define*

$$F \vdash G \overset{def}{\Longleftrightarrow} \mathbf{Upd}(KB, F) \vdash G \text{ for all } KB \in L_{\text{KB}}$$

More general, for finite $\Phi \subseteq L_{\text{Input}}$, $\Phi \vdash G$ is defined as $\bigwedge \Phi \vdash G$.

The generality of the so-defined (finite) entailment relation does not depend on L_{KB}. Notice that the standard definition of entailment (compare with $\models$ $F \overset{def}{\Longleftrightarrow} \emptyset \models F$) corresponds to

$$F \vdash G \overset{def}{\Longleftrightarrow} \mathbf{Upd}(0, F) \vdash G,$$

which is equivalent to the general definiton if the KRS satisfies Monotonicity and Permutation, and KBs are input-constructible.

Dynamic Semantics

In dynamic semantics, such as the *belief models* of [Gärdenfors 1984+1988], or the *update semantics* of [Veltman 1990], the meaning of a sentence consists in the change it brings about in the information state of an agent who accepts the news conveyed by it. In other words, sentences F are considered as update functions

$$F : L_{\text{KB}} \rightarrow L_{\text{KB}}, \quad \text{where } F(KB) := \mathbf{Upd}(KB, F)$$

Other concepts, like logical implication and equivalence, are derivative, that is, definable in terms of updates.

In these theories a KB is usually a unique representation of the information contained in it (for instance, a set of two-valued models, or a deductively closed set of formulas). Therefore, the inference relation can be defined according to the intuitive requirement that F should be inferrable from an information state KB if the information expressed by F does not contain anything 'new' for KB, i.e.

$$\text{KB} \vdash F \overset{\text{def}}{\Longleftrightarrow} \textbf{Upd}(\text{KB}, F) = \text{KB}$$

However, in terms of logical generality and computational efficiency this is certainly not a good definition of inference. Since it implies Cumulativity, it is only interesting as a particular way of obtaining the logic of a cumulative KRS from one fundamental concept, viz. the update operation.

In Veltman's update semantics a KB consists of sets of atoms, $\text{KB} \subseteq 2^{\text{At}}$, and is called an *information state*. It contains those 'worlds' which – for all is known – may yet turn out the real one. As an agent's knowledge increases, KB shrinks. The informationally empty KB is the set of all subsets of At, $0 = 2^{\text{At}}$ The update operation is defined as

(a) $\textbf{Upd}(\text{KB}, a) = \{w \in \text{KB} : a \in w\}$
$(\neg)$ $\textbf{Upd}(\text{KB}, \neg F) = \text{KB} - \textbf{Upd}(\text{KB}, F)$
$(\wedge)$ $\textbf{Upd}(\text{KB}, F \wedge G) = \textbf{Upd}(\text{KB}, F) \cap \textbf{Upd}(\text{KB}, G)$
$(\vee)$ $\textbf{Upd}(\text{KB}, F \vee G) = \textbf{Upd}(\text{KB}, F) \cup \textbf{Upd}(\text{KB}, G)$

Veltman defines logical consequence with respect to minimal information:

$$F \models G \overset{\text{def}}{\Longleftrightarrow} \textbf{Upd}(0, F) \vdash G$$

In this way he, in fact, gets classical logic. The *proposition* $\lceil F \rceil$ expressed by the sentence F is defined as $\lceil F \rceil := \textbf{Upd}(0, F)$. It holds, then, that

$$\textbf{Upd}(\text{KB}, F) = \text{KB} \cap \lceil F \rceil$$

Observation 2.6 (Belief Revision)   *Base revision functions in the theory of belief revision can be viewed as special cases of update operations. The following postulates are adapted from analogues in [Gärdenfors 1988] where they are formulated for expansion and contraction.*

(Success) $\textbf{Upd}(\text{KB}, F) \vdash F$
(Vacuity) $\text{KB} \vdash F \Rightarrow \textbf{Upd}(\text{KB}, F) \cong \text{KB}$
(Extensionality) $F \vdash G \ \& \ G \vdash F \Rightarrow \textbf{Upd}(\text{KB}, F) \cong \textbf{Upd}(\text{KB}, G)$

Instead of equality, equivalence (defined below) is used in the formulation of Vacuity and Extensionality, since KBs may not be unique representations. The postulate of Success is nothing else as unrestricted reflexivity, whereas Vacuity is slightly stronger than Cumulativity. Another postulate, called Inclusion *in the AGM theory, expresses the property of updates being ampliative in the case of expansion, and reductive in the case of contraction.*

Extensionality follows from Lemma Redundancy and by the above definition of $F \vdash G$. Notice, however, that the AGM theory[27] of belief revision relies heavily on classical logic, and therefore Extensionality is a kind of soundness condition with respect to classical logic, whereas here every KRS defines its own logic (by its associated inference relation).

2.7   Rule Knowledge Representation Systems

With each KRS K a *rule-based* extension, RK, which is also called *rule knowledge representation and reasoning system (RKRS)*, can be associated. In RK, a knowledge base $X \in L_{KB}$ is supplemented by a set $R \subseteq L_{Input} \times L_{Query}$ containing rules $r = \langle Conclusion, Premise \rangle$ with $Conclusion \in L_{Input}$ and $Premise \in L_{Query}$, also written as '*Conclusion $\leftarrow$ Premise*'. These rules are mappings between KBs,[28]

$$r : L_{KB} \rightarrow L_{KB}$$

since – in the standard case[29] – their application is defined as

$$r(X) = \begin{cases} \mathbf{Upd}(X, Conclusion) & \text{if } X \vdash Premise \\ X & \text{otherwise} \end{cases}$$

Observation 2.7   *If the KB X of a rule knowledge base $\langle X, R \rangle$ consists of a set of compiled input formulas $Y \subseteq L_{Input}$, $X = \lceil Y \rceil$, then the rule knowledge base can be rewritten as an equivalent set of rules:*

$$\langle X, R \rangle \quad \cong \quad \langle 0, R \cup \{F \leftarrow 1 : F \in Y\} \rangle$$

where those 'improper' rules with a trivially true premise, $F \leftarrow 1$, represent the 'facts'.

---

[27] Called after its originators, C. Alchourrón, P. Gärdenfors and D. Makinson.   See [Gärdenfors 1988].

[28] The interpretation of rules as mappings proposed here amounts to an *operational semantics* which is also advocated by Belnap [1977]. The operational approach to rules, though in a less formal way, seems to be common in many AI systems where a mechanism for 'firing', resp. 'triggering', rules is provided. Alternatively, and more in line with a model-theoretic semantics, one can define a rule as a pair of query formulas $ConclusionQuery \leftarrow PremiseQuery$ such that it constrains the possible closures $Z \in R(X)$: $Z \vdash ConclusionQuery$ whenever $Z \vdash PremiseQuery$. This could be called a *constraint semantics* of rules, since they are read as a kind of constraints on closures (resp. models).

[29] Refined methods of rule application for defeasible reasoning are discussed, e.g., in [Wagner 1991c].

Definition 2.6   *A mapping $f : A \to A$ from a preorder $\langle A, \leq \rangle$ into itself is called* monotonic *if $f(x) \leq f(y)$ whenever $x \leq y$. It is called* ampliative *if $x \leq f(x)$. A rule is called* monotonic *(resp.* ampliative*) if it is a monotonic (resp. ampliative) mapping.*

Observation 2.8   *The rule $F \leftarrow G$ is monotonic if its conclusion F is ampliative, and its premise G is persistent.*

The semantics of a rule knowledge base $\langle X, R \rangle$ is determined by the definition of a preferred closure of X under R, being a knowledge base $Z \in L_{\mathrm{KB}}$ closed under all rules of R:

$$r(Z) = Z \quad \text{for all } r \in R$$

In general, however, there may be several preferred closures, or none. Their collection is denoted by $R(X)$ which is identified with its only element if it is a singleton. If there are several preferred closures, a valid consequence must be inferrable from all of them:

$$C(\langle X, R \rangle) \quad := \quad \bigcap \{ C(Z) : Z \in R(X) \}$$

In the simplest case, namely for monotonic KRSs, the informationally minimal closures extending X are the preferred ones (yielding unique preferred closures which are computable by successively applying all rules in an arbitrary order in the case of definite KBs, such as normal or extended logic programs without negation-as-failure). In the case of nonmonotonic ampliative KRSs, such as normal or extended logic programs, not all minimal closures are preferred. If the rule set R is, in some sense, *wellfounded*, then there is a 'natural' stratification of R determining the right order of rule application such that a unique preferred closure is obtained as the result of closing X by successively applying all rules $r \in R$ in that order.

Determining the preferred closures is especially difficult if the dependency graph of R contains loops involving weak negation, or in other words, if R is not weakly wellfounded (see 7.5). In this case the concept of a *stable closure* generalising the stable model semantics of [Gelfond & Lifschitz 1988] offers help.

By means of stable closures, a general form of negation-as-failure, called *stable negation*, can be added to the premises of rules of any ampliative RKRS provided that it has unique closures $R(X)$ for any rule knowledge base $\langle X, R \rangle$. Instead of rules $F \leftarrow G$ we allow for rules

$$F \leftarrow G_1, \ldots, G_i, \mathbf{not}\, G_{i+1}, \ldots, \mathbf{not}\, G_m$$

where **not** is a newly introduced operator (not occuring in the query and input language, that is, F, G and all G_k do not contain **not**).

Definition 2.7 (Stable Closure)   *For any RKB $\langle X, R \rangle$ and any $Z \in L_{KB}$ define R^Z as the set of not-free rules obtained from R by*

1. *removing all rules containing in the premise not G such that $Z \vdash G$, and*

2. *deleting all remaining expressions not G from the premise of rules.*

$Z \in L_{KB}$ is called a stable closure of $\langle X, R \rangle$ if $Z = R^Z(X)$.

Example 2.5   *Normal logic programs correspond to rule knoweldge bases of RV_{RDB} being ampliative but not persistent. For $X = \{p(c)\}$ and the wellfounded*

$$R_1 = \{q(x) \leftarrow \text{not } r(x)\}$$

we obtain the unique preferred closure

$$R_1(X) = \{p(c), q(c)\}$$

while for the non-wellfounded

$$R_2 = \{q(x) \leftarrow \text{not } r(x), r(x) \leftarrow \text{not } q(x)\}$$

we obtain

$$R_2(X) = \{\{p(c), q(c)\}, \{p(c), r(c)\}\}$$

according to the stable closure semantics.

2.8   Equivalence between KBs and KRSs

The concept of a KRS consisting of an answer and an update operation suggests two notions of equivalence[30] between two KBs of a KRS.

Definition 2.8 (Answer Equivalence)   KB$_1$ *and* KB$_2$ *are answer equivalent, symbolically* KB$_1 \simeq$ KB$_2$, *if they produce the same answers at the moment, i.e.*

$$\textbf{Ans}(\text{KB}_1, F) = \textbf{Ans}(\text{KB}_2, F)$$

for any query formula F.

Definition 2.9 (Update Equivalence)   KB$_1$ *and* KB$_2$ *are update equivalent, symbolically* KB$_1 \cong$ KB$_2$, *if they produce the same answers in all circumstances, i.e.*

$$\textbf{Upd}(\text{KB}_1, F) \simeq \textbf{Upd}(\text{KB}_2, F)$$

for any input formula F.

---

[30]Belnap [1977] proposes essentially the same notions in his framework of *information states* calling them *current* and *strong* equivalence.

Finally, two (input-constructible) KRSs are equivalent if their resp. empty KBs are update equivalent.

Definition 2.10   *The KRSs K and K' are equivalent, symbolically $K \simeq K'$, if*

1. $L_{\text{Query}} = L'_{\text{Query}}$, *and*

2. $L_{\text{Input}} = L'_{\text{Input}}$, *and*

3. $L_{\text{Answer}} = L'_{\text{Answer}}$, *and*

4. *for all $F \in L_{\text{Input}}$ and all $G \in L_{\text{Query}}$,*

$$\mathbf{Ans}(\mathbf{Upd}(0, F), G) = \mathbf{Ans}'(\mathbf{Upd}'(0, F), G)$$

Example 2.6   *Let $X \subseteq 2^{\text{At}}$, and $F \in L(\wedge, \vee)$, and*

$$\mathbf{Upd}(X, F) := \{A \cup B : A \in X \ \& \ B \in \text{DNS}(F)\}$$

and

$$X \vdash F \overset{def}{\Longleftrightarrow} \forall A \in X \ \exists B \in \text{DNS}(F) : B \subseteq A$$

where DNS denotes the disjunctive normal set operator defined in 3.1.5. Then,

$$\langle 2^{2^{\text{At}}}, \vdash, L(\wedge, \vee), \mathbf{Upd}, L(\wedge, \vee) \rangle$$

$$\simeq \quad \langle 2^{L(\wedge, \vee)}, \vdash_P, L(\wedge, \vee), \cup, L(\wedge, \vee) \rangle$$

$$\simeq \quad \langle 2^{L(\wedge, \vee)}, \vdash_H, L(\wedge, \vee), \cup, L(\wedge, \vee) \rangle$$

$$\simeq \quad \langle 2^{L(\wedge, \vee)}, \vdash_{cl}, L(\wedge, \vee), \cup, L(\wedge, \vee) \rangle$$

where $\vdash_P$ denotes positive logic (see 3.3.2), $\vdash_H$ denotes Heyting's intuitionistic logic, and $\vdash_{cl}$ denotes classical logic.

Example 2.7   *Classical and three-valued logic aggree on literal consequences (see 3.1.6):*

$$\langle 2^{L(\sim, \wedge, \vee)}, \models_2, \text{Lit}, \cup, L(\sim, \wedge, \vee) \rangle \quad \simeq \quad \langle 2^{L(\sim, \wedge, \vee)}, \models_3, \text{Lit}, \cup, L(\sim, \wedge, \vee) \rangle$$

Example 2.8   *Let V^+_{RDB} be equal to V_{RDB} except that the query language now does not contain weak negation, i.e. only positive queries are allowed, $L_{\text{Query}} = L(\wedge, \vee)$. Let*

$$V^+_{\text{LP}} \quad := \quad \langle 2^{\{a \leftarrow A : a \in \text{At} \supseteq A\}}, \vdash, L(\wedge, \vee), \cup, \text{At} \rangle$$

denote the system of positive logic programs. Then

$$RV^+_{\text{RDB}} \simeq V^+_{\text{LP}}$$

i.e. the rule-based extension of relational databases is equivalent to the system of positive logic programs.

2.9   Levesque's Concept of a Vivid Knowledge Base

According to Levesque [1986], for vivid knowledge "there will be a one-to-one correspondence between entities and relationships of interest in the world and certain symbols and appropriate connections between them in the vivid knowledge base (VKB)". He calls a KB 'incomplete' if "it tells us that one of a number of sentences is true but doesn't tell us which". As the sources of 'incompleteness' he lists existential, universal, disjunctive and negated sentences. 'Complete' knowledge, according to Levesque, is decomposable into atomic knowledge. He therefore suggests [1988] to allow in a VKB only ground atoms, called *vivid facts*, satisfying the above vividness criterion. The semantics of such a VKB would be given by the domain closure axiom (stating that only those things which are explicitly mentioned in the KB exist), the CWA and the usual equality axioms. The VKB would be consistent and complete (VKB $\vdash F$ or VKB $\vdash -F$). Query answering ultimately reduces to table lookup on atomic sentences.

Apparently, Levesque [1986] considers reasoning with the CWA and reasoning without any form of CWA as mutually exclusive. I will argue, however, that it is possible, and moreover, that it makes sense to assume the CWA for certain predicates while assuming the 'Open-World Assumption' for others. Following the general idea of vivid reasoning the goal is to establish a logic along the lines of two basic principles: cognitive adequacy and computational feasibility. These principles, clearly, are more relevant to AI than any philosophical dogma like that of bivalence constituting classical logic. So we should not only be willing to deviate from classical logic but also actively seek for deviations which make sense for our enterprise.

Levesque gives an example of a KB that is complete but not vivid, namely

> *Jack and Dan together polished off a 13 ounce bottle of gin.*
> *Dan had precisely one more drink (1 ounce) than Jack.*

Its vivification yields

> *Dan drank exactly 7 ounces of gin.*
> *Jack drank exactly 6 ounces of gin.*

In [Etherington et al. 1989] it is argued that a "VKB cannot explicitly represent disjunction, negation, or any form of incompleteness." and that "Disjunctive and negative information do not fit readily into the database world-view, and are major contributors to the complexity of logical reasoning."

Clearly, disjunctive and existential information is mainly responsible for 'incompleteness'. It is not quite clear in which respect universal sentences represent 'incomplete' information since logic programming rules are (implicitly) universally quantified and this does not create any problem.

It is, however, the case of negated sentences where I mainly disagree with the above account of 'incompleteness'. Explicit negative information in the form of (strongly) negated sentences plays an important role in natural discourse and is certainly not 'incomplete' in the same sense as disjunctive information.[31] Levesque might not have seen the ambiguity of natural negation being strong (and definite) in some contexts, and being weak (and 'incomplete') in others.

The problem of disjunctive information is illustrated in [Levesque 1986] with quite impressive figures on the computational requirements needed to handle disjunctive facts like

$$married(Jack, Jill) \lor married(Jack, Jan).$$

Imagine a party with, say, two hundred guests, where your task is to figure out an appropriate seating arrangement satisfying a number of constraints. If all you know are 25 such disjunctive 'married' facts, you would have to form every possible combination of atoms implied by these disjunctions, that is, check 30 million possibilities which would take 30 seconds on a machine looking at a million possibilities a second. So, reasoning with 25 such sentences is still feasible. Now suppose you get 100 disjunctive facts (which does not seem to be totally unrealistic in this setting if you are willing to accept this kind of information). There is no way to process this disjunctive knowledge base: even if you employ a machine with a speed-up of one million (i.e. looking at a million million possibilities a second which is way beyond current technology) it would take 30 billion years !

Levesque [1988] suggests one possible solution to avoid disjunctive information (and the combinatorial explosion caused by it): instead of adding a disjunction to the KB, choose a common upper bound of the disjuncts, subsuming each one of them, and add it as an approximation of the disjunction to the KB. In [Etherington et al. 1989] it is proposed to use a taxonomy, or type hierarchy, in order to find an appropriate superclass. For instance, the disjunctive fact

$$lecturer(Peter) \lor assistant\_professor(Peter)$$

can be approximated by the definite fact

$$university\_teacher(Peter)$$

---

[31] A nice example of a vivid information system where both positive and explicit negative information occur naturally is discussed in [Pearce 1991]: "When parking a car, for instance, typically the information available is negative, in the form of prohibitions. Road signs convey information that parking is forbidden, or restricted to certain times, vehicles or places. In, and only in, the absence of (negative) information to the contrary do we assume we are legally permitted to park in a given spot. In this situation, permission is the default [...] Actually, road use is an example of a 'mixed information' system. In many of today's congested cities, prohibition is the parking default. In that case, I need hardly waste time examining the restriction signs. Much simpler to look for a large $\boxed{P}$ indicating a car-park or unrestricted zone."

together with the subsumption rules

$$university\_teacher(x) \leftarrow lecturer(x)$$
$$university\_teacher(x) \leftarrow assistant\_professor(x)$$

Such an approximation may not always be possible, especially, in the case of indefinite individuals, like e.g.

$$lecturer(Peter) \vee lecturer(Tom).$$

So, it remains an open question how to accommodate disjunctive information in a KB without risking combinatorial explosion.

Question 2   *Do we need a specific mechanism in order to restrict the amount of disjunctive information in a KB ?*

Several aspects of Levesque's concept of a VKB conflict with my considerations of vividness. First, the domain closure axiom violates the WYSIWYM principle: by collecting information in a VKB a cognitive agent does certainly not assume that his universe of discourse is closed and constant. On the contrary, it seems to be natural that the universe changes dyamically as new individuals become known to the agent. Another WYSIWYM violation is the CWA, if it is meant to relate default-implicit negative information with classical negation for all predicates. Second, consistency is not essential for a VKB. On the contrary, a vivid KRS should be able to tolerate inconsistency, as I have argued above.

2.10   On the Concept of a VKRS

Depending on the specific requirements of an application, and on the computational resources available, different answer and update operations constituting different KRS's may be appropriate. However, there are some minimal requirements any vivid KRS has to satisfy:

1. Restricted Reflexivity

2. Constructivity

3. Non-Explosiveness

In the sequel, a KRS that satisfies these conditions is called a *basic VKRS*.

Additionally, besides a principal negation (called *strong*), expressing explicit falsity, there should be a second negation (called *weak*) which handles default-implicit negative information in the style of negation-as-failure. Thus, the query language (and possibly also the input language) should contain $\sim$ and $-$.

Some form of CWA, restricted to specific predicates (namely those which are totally represented in the knowledge base) then relates explicit with default-implicit falsity, i.e. strong with weak negation: an atomic sentence formed with a totally represented predicate is explicitly false if it is implicitly false by default, i.e. its strong negation holds if its weak negation does.

Definition 2.11 (CWA)   *In order to handle total predicates in a VKRS, a KB has to define which predicates are positive-totally and which ones are negative-totally represented, for example by specifying two respective sets CWA^+ and CWA^-, and applying them in the derivation of literals:*

$(CWA+)$ $\text{KB} \vdash \sim p(t)$ *if* $p \in CWA^+$ & $\text{KB} \vdash -p(t)$
$(CWA-)$ $\text{KB} \vdash p(t)$ *if* $p \in CWA^-$ & $\text{KB} \vdash -\sim p(t)$

Definition 2.12 (VKB)   *A vivid KB, in its general form, is a triple*

$\langle CWA, KB, IC \rangle$

where $CWA = \langle CWA^+, CWA^- \rangle$ specifies the predicates subject to the Closed-World Assumption, and $IC \subseteq L_{\text{Query}}$ is a set of integrity constraints.

We shall in most cases neglect integrity constraints in this paper. In cases where CWA^- is empty, we identify CWA with CWA^+. If neither the CWA nor integrity constraints matter, we only mention KB.

Definition 2.13 (VKRS)   *A basic VKRS extended by adding weak negation and the above CWA rules is called a (full) VKRS.*

If a VKRS enjoys the property of Lemma Compatibility, it will be called *cumulative*. In certain respects, a cumulative VKRS ispreferable to a non-cumulative one since it is more regular. However, there seems to be a trade-off between regularity and the capability to capture all aspects of commonsense reasoning. In particular, the price for cumulativity seems to be a more cautious inference operation not allowing for certain inferences justified by respective commonsense reasoning schemes.

We will use the following notation for a KRS: let K denote a KRS where only the query but not the input language contains weak negation, then K^+ denotes its weak-negation-free fragment not allowing for weak negation in queries, and K^- denotes its extension by adding weak negation to the input language.

2.10.1 Simplification of Compound Formulas

In a VKRS we have the following DeMorgan-like rewrite rules in order to simplify compound formulas:

$$-(F \wedge G) \longrightarrow -F \vee -G$$

$$
\begin{aligned}
\sim(F \wedge G) &\longrightarrow \sim F \vee \sim G \\
-\sim(F \wedge G) &\longrightarrow -\sim F \wedge -\sim G \\
-(F \vee G) &\longrightarrow -F \wedge -G \\
\sim(F \vee G) &\longrightarrow \sim F \wedge \sim G \\
-\sim(F \vee G) &\longrightarrow -\sim F \vee -\sim G \\
\sim\sim F &\longrightarrow F \\
-\sim\sim F &\longrightarrow -F \\
\sim -F &\longrightarrow F \\
--F &\longrightarrow F \\
-\sim -F &\longrightarrow -F
\end{aligned}
$$

For inductive definitions, then, it is sufficient to treat the cases of the verum (1), of extended literals (e), of conjunctions ($\wedge$), and of disjunctions ($\vee$). All other cases are covered by the above rewrite rules which are justified in 3.1 (observation 3.4).

2.10.2   How to Construct a Definite VKRS

If only definite knowledge is to be represented, there is a natural way of defining derivability and constructing a VKRS. It can be sketched as follows:

1. Stipulate the data structures to be used for knowledge representation, i.e. L_{Repr}, resp. L_{KB}.

2. Define derivability for ground literals, $\text{KB} \vdash l$, and weakly negated ground literals, $\text{KB} \vdash -l$, yielding an inference relation $\vdash \subseteq L_{\text{KB}} \times \text{XLit}$, such that Inherent Consistency for weak negation holds:

$$\text{KB} \vdash e \;\Rightarrow\; \text{KB} \not\vdash \bar{e},$$

   When a definite KB is based on a finite (Herbrand) universe, it even holds that $\text{KB} \vdash e$ iff $\text{KB} \not\vdash \bar{e}$.

3. Extend the relation obtained in step 2 to an inference relation between KBs and arbitrary ground formulas $F \in L(1, -, \sim, \wedge, \vee)$ with the help of the above rewrite rules and the following inductive clauses:

   (1) $\text{KB} \vdash 1$
   ($\wedge$) $\text{KB} \vdash F \wedge G$ if $\text{KB} \vdash F$ and $\text{KB} \vdash G$
   ($\vee$) $\text{KB} \vdash F \vee G$ if $\text{KB} \vdash F$ or $\text{KB} \vdash G$

These rules correspond to introduction rules of a natural-deduction-like system. The respective elimination rules will have to be taken into account in the definition of an appropriate update operation.

Disjunction is definable by means of the DeMorgan identity.

$$F \vee G = \sim (\sim F \wedge \sim G)$$

Thus, in order to establish inductive definitions and proofs it is not necessary to consider disjunction.

4. Define an answer operation for ground queries based on the inference relation $\vdash$.

5. Define the update operation for (extended) ground literals in such a way that it satisfies Contraction, Permutation and Restricted Reflexivity. Stipulate that

    (1) $\mathbf{Upd}(KB, 1) = KB$

For weakly consistent inputs (see below) we have

    $(U\wedge)$ $\mathbf{Upd}(KB, F \wedge G) = \mathbf{Upd}(\mathbf{Upd}(KB, F), G)$

From Restricted Reflexivity and Inherent Consistency follows that

    $\mathbf{Upd}(KB, -F) \not\vdash F$

Thus, updating KB with $-F$ amounts to what is called the *contraction* of KB by F, symbolically $KB \div F$, in the literature on belief revision (see e.g. [Gärdenfors 1988]). I will not pursue this point further here but only mention some advantages of using the logical operator '$-$' instead of the metalogical '$\div$'. First, the 'deletion interpretation' of weakly negated inputs seems to be an interesting generalization of negation-as-failure which is usually restricted to query formulas only[32]. Second, as I learned from André Fuhrmann [1992], there is a discussion in philosophical logic on *logical subtraction* dating back to Wittgenstein's "Tractatus". It seems that the subtraction of G from F, in Fuhrmann's notation '$F - G$', can be expressed in vivid logic as $F \wedge -G$.

6. If wanted, extend the answer operation by allowing for

    (a) nonground query formulas yielding answer relations, i.e. sets of substitution tuples,

    (b) conditional and other intensional query formulas, as described in 2.6.2.

---

[32] A notable exception is [Lifschitz & Woo 1992].

Chapter 3

Partiality, Paraconsistency and Constructivity

In classical logic sentences are either true or false, and it is assumed that at any time every sentence has exactly one of these two truth-values regardless of the available information about it. In a KRS, however, information is rather partial, and may be even incoherent. Therefore, partial logic is presented as a model for reasoning with partial and possibly inconsistent information.

Belnap's concept of an epistemic state, resp. information state, is based on partial logic. It is shown that Belnap's systems are fundamental cases of basic VKRSs.

The principle of constructivity does not only apply to disjunction and conjunction, but also to implication. Constructive implication is presented as the proper logical operator expressing the metalogical conditional connective of rules in the object language.

3.1 Partial Logic

There has been a growing interest in partial logics over recent years, mainly stimulated by problems in formal linguistic and artificial intelligence. One important area of application (which is also essential to the work presented here) is knowledge representation and nonmonotonic reasoning. The former is addressed, e.g., in [Thijsse 1992] where besides a broad discussion of metalogical issues knowledge representation is treated in the tradition of epistemic logics. The latter is adressed, e.g., in [Doherty 1991] and [Tan 1992] where two interesting formalisms of nonmonotonic reasoning based on partial semantics are presented.

I will only treat here those parts of partial logic that are needed in order to formulate and understand the proposed framework of vivid knowledge representation and reasoning.

3.1.1 Partial Models

Let $\mathcal{M} = \langle M^t, M^f \rangle$ be a partial Herbrand interpretation, that is, M^t contains the positive facts which are believed to be true, whereas M^f contains the negative facts which are believed to be false (formally, $M^t, M^f \subseteq \text{At}$). Following Langholm [1988], I will speak of *proper*, or *coherent*, interpretations when M^t and M^f are required to be disjoint, as opposed to *general* interpretations for which they may overlap (in the sequel, I will frequently just say 'interpretation', or 'model', instead of 'general partial interpretation', resp. 'model').

A partial Herbrand interpretation gives rise to a model as well as to a countermodel relation, inductively defined as follows:

$$
\begin{aligned}
\mathcal{M} &\models a & \text{iff} \quad & a \in M^t \\
\mathcal{M} &\models F \wedge G & \text{iff} \quad & \mathcal{M} \models F \text{ and } \mathcal{M} \models G \\
\mathcal{M} &\models F \vee G & \text{iff} \quad & \mathcal{M} \models F \text{ or } \mathcal{M} \models G \\
\mathcal{M} &\models -F & \text{iff} \quad & \mathcal{M} \not\models F \\[6pt]
\mathcal{M} &\models \sim F & \text{iff} \quad & \mathcal{M} \mathrel{=\!\mid} F \\
\mathcal{M} &\mathrel{=\!\mid} a & \text{iff} \quad & a \in M^f \\
\mathcal{M} &\mathrel{=\!\mid} F \wedge G & \text{iff} \quad & \mathcal{M} \mathrel{=\!\mid} F \text{ or } \mathcal{M} \mathrel{=\!\mid} G \\
\mathcal{M} &\mathrel{=\!\mid} F \vee G & \text{iff} \quad & \mathcal{M} \mathrel{=\!\mid} F \text{ and } \mathcal{M} \mathrel{=\!\mid} G \\
\mathcal{M} &\mathrel{=\!\mid} \sim F & \text{iff} \quad & \mathcal{M} \models F \\
\mathcal{M} &\mathrel{=\!\mid} -F & \text{iff} \quad & \mathcal{M} \models F
\end{aligned}
$$

where a, F and G are ground, and the model relation holds for an open formula if it holds for some ground instance of it,[1]

$$\mathcal{M} \models F(x) \quad \text{iff} \quad \mathcal{M} \models F(t) \quad \text{for some ground term } t$$

Also, it is stipulated that for all interpretations $\mathcal{M}$, $\mathcal{M} \models 1$. There are two notions of falsity involved in this semantics each one underlying the respective negation: $\mathcal{M} \mathrel{=\!\mid} F$ stands for the explicit falsity, resp. falsifiability, of F, whereas $\mathcal{M} \not\models F$ stands for the weak falsity, resp. non-verifiabilty, of F given implicitly. The dual notion of weak falsity, 'weak truth' alias $\neq\!\mid$, is equivalent to the weak faslity of $\sim F$: $\mathcal{M} \neq\!\mid F$ iff $\mathcal{M} \models -\sim F$. So we have four basic semantical modes of a formula F, its truth, F, its falsity, $\sim F$, its weak falsity, $-F$, and its weak truth, $-\sim F$. Truth and falsity are independent from each other and behave dually symmetric.

Note that the first four clauses of the model definition correspond to classical logic where $\mathcal{M}$ would be *coherent* and *total*, i.e. $M^t \cap M^f = \emptyset$ and $M^t \cup M^f = \text{At}$,

---

[1] Notice that throughout this book, if it is not essential, I will not bother about arity. Instead of, e.g., "$p(t_1, \ldots, t_n)$" for a n-place predicate p and ground terms $t_1, \ldots, t_n$, I will simply write "$p(t)$" and leave the full generalization to the reader.

and consequently, falsity and weak falsity, resp. truth and weak truth, would coincide.

Observation 3.1   *Let $\mathcal{M}$ be coherent and total. Then*

   (i) $\mathcal{M} \models \sim F$ *iff* $\mathcal{M} \models -F$.

   (ii) $\mathcal{M} \models F$ *iff* $\mathcal{M} \models -\sim F$.

Proof (by induction on F):

$\underline{F = a}$ (i) If $\mathcal{M} \models \sim a$ then $a \in M^f$, and so by coherency, $a \notin M^t$, i.e. $\mathcal{M} \models -a$. Conversely, if $\mathcal{M} \models -a$ then $a \notin M^t$, thus by totality, $a \in M^f$, i.e. $\mathcal{M} \models \sim a$. (ii) Dual symmetrically to (i).

$\underline{\wedge, \vee}$ For conjunctions and disjunctions the assertions follow by straightforward induction.

$\underline{F = -G}$ (i) and (ii) follow immediately from the definition of the model relation.

$\underline{F = \sim G}$ (i) $\mathcal{M} \models \sim\sim G$ iff $\mathcal{M} \models G$. By the induction hypothesis for (ii) this is equivalent to $\mathcal{M} \not\models G$, respectively $\mathcal{M} \models -\sim G$. (ii) By the induction hypothesis for (i), $\mathcal{M} \models \sim G$ iff $\mathcal{M} \models -G$, which is equivalent to $\mathcal{M} \models -\sim\sim G$. □

The model-theoretic consequence relation between a set of premise formulas X and a conclusion formula F, $X \models_n F, (n = 2, 3, 4)$, is defined as usual: it holds if every n-valued model of X is also a model of F. 2-valued, resp. 3-, resp. 4-valued, consequence refers to classical (i.e. total and coherent), resp. proper, resp. general partial models.

Since there is no need for it, I will not define here the countermodel relation for non-ground formulas. Conjunction and disjunction are interdefinable in the usual way, i.e. by the DeMorgan identities (relating disjunction with conjunction and strong negation), so definitions and proofs can be restricted to the cases without $\vee$.

In order to simplify notation an interpretation $\mathcal{M}$ is also represented as the set M of all ground literals supported by it,

$$M = \{l \in \mathrm{Lit} : \mathcal{M} \models l\}$$

M is also called the *diagram* of $\mathcal{M}$.

$\mathcal{M}'$ is called an *extension* of $\mathcal{M}$, symbolically $\mathcal{M}' \geq \mathcal{M}$, if $M \subseteq M'$. An extension of a model represents a growth of information since it assigns truth or falsity to formerly undetermined sentences.

Definition 3.1   *Let $\mathcal{M}' \geq \mathcal{M}$. Then a formula F is called* persistent *in partial logic if*

(i) $\mathcal{M} \models F \Rightarrow \mathcal{M}' \models F$

(ii) $\mathcal{M} =\!| F \Rightarrow \mathcal{M}' =\!| F$

Persistent formulas remain true, resp. false, when more information becomes available. Not all formulas are persistent.

Observation 3.2   *For instance, $-q$ is not persistent: $\{p\} \models -q$, but $\{p, q\} \not\models -q$. However, formulas without weak negation are persistent.*

Proof by induction on the complexity of F:

$\underline{F = a}$ Then $a \in M$ whenever $\mathcal{M} \models a$. Consequently, $a \in M' \supseteq M$, i.e. $\mathcal{M}' \models a$. Similarily, $\sim\!a \in M$ whenever $\mathcal{M} =\!| a$, implying that $\sim\!a \in M'$, i.e. $\mathcal{M}' =\!| a$.

$\underline{F = G \wedge H}$ If $\mathcal{M} \models G \wedge H$ then $\mathcal{M} \models G$ and $\mathcal{M} \models H$, consequently by the induction hypothesis, $\mathcal{M}' \models G$ and $\mathcal{M}' \models H$, implying $\mathcal{M}' \models G \wedge H$. On the other hand, if $\mathcal{M} =\!| G \wedge H$ then $\mathcal{M} =\!| G$ or $\mathcal{M} =\!| H$. Consequently, by the induction hypothesis, $\mathcal{M}' =\!| G$ or $\mathcal{M}' =\!| H$, so in any case $\mathcal{M}' =\!| G \wedge H$.

$\underline{F = \sim\!G}$ If $\mathcal{M} \models \sim\!G$ then $\mathcal{M} =\!| G$. Therefore, by the induction hypothesis, $\mathcal{M}' =\!| G$, implying that $\mathcal{M}' \models \sim\!G$. On the other hand, if $\mathcal{M} =\!| \sim\!G$ then $\mathcal{M} \models G$ and, by the induction hypothesis, also $\mathcal{M}' \models G$, implying that $\mathcal{M}' =\!| \sim\!G$.

The fact that formulas without weak negation are persistent is called the *permanence principle* in [Körner 1966]. It can also be regarded as expressing the monotonicity of truth and falsity with respect to growing information.

Definition 3.2 (Minimal Model) $\mathcal{M}$ *is called a* minimal model *of F, symbolically $\mathcal{M} \in MinMod(F)$, if*

(i) $\mathcal{M} \models F$, *and*

(ii) *If $\mathcal{M}' \models F$, and $\mathcal{M}' \leq \mathcal{M}$, then $\mathcal{M}' = \mathcal{M}$.*

Observation 3.3   *If restricted to persistent formulas, $F \in L(1, \sim, \wedge, \vee) \supseteq X$, the model-theoretic consequence relation in partial semantics is completely determined by the minimal models:*

$$X \models_4 F \quad \textit{iff} \quad \mathcal{M} \models F \quad \textit{for all } \mathcal{M} \in MinMod(X)$$

Observation 3.4   *As can be easily checked, for any of the DeMorgan-like rewrite rules of the form $LHS \rightarrow RHS$ listed in 2.10, it holds that*

$$\mathcal{M} \models LHS \quad \textit{iff} \quad \mathcal{M} \models RHS$$

whereas

$$M \dashv LHS \quad \textit{iff} \quad M \dashv RHS$$

holds in all cases except $(\sim -)$ *and* $(--)$.

3.1.2 Four-Valued Truth Tables

A model $\mathcal{M}$ determines a four-valued assignment v_M on ground atoms a in the following way:

$$v_M(a) = \left\{ \begin{array}{c} 1 \\ 0 \\ u \\ o \end{array} \right\} \quad \text{if} \quad \{a, \sim a\} \cap M = \left\{ \begin{array}{c} \{a\} \\ \{\sim a\} \\ \{\} \\ \{a, \sim a\} \end{array} \right.$$

where $1, 0, u$ and o stand for *true, false, undetermined* and *overdetermined*.

The recursive truth and falsity definitions above correspond to the following truth tables:

$\wedge$	0	u	o	1
0	0	0	0	0
u	0	u	0	u
o	0	0	o	o
1	0	u	o	1

$\vee$	0	u	o	1
0	0	u	o	1
u	u	u	1	1
o	o	1	o	1
1	1	1	1	1

$\sim$	
0	1
u	u
o	o
1	0

$-$	
0	1
u	1
o	0
1	0

Observation 3.5   *If $\bar{v}_M$ is the valuation extending v_M according to these tables in the usual way, i.e. as a homomorphism from the free algebra of formulas into the algebra $\langle\{0, u, o, 1\}, \wedge, \vee, \sim, -\rangle$, we obtain*

$$\bar{v}_M(F) = \left\{ \begin{array}{llll} 1 & \textit{if} & M \models F \ \& \ M \not\dashv F \\ 0 & \textit{if} & M \dashv F \ \& \ M \not\models F \\ u & \textit{if} & M \not\models F \ \& \ M \not\dashv F \\ o & \textit{if} & M \models F \ \& \ M \dashv F \end{array} \right.$$

The connectives $\wedge, \vee, \sim$ given by the corresponding 3-valued versions of these tables were first investigated by Lukasiewicz [1920] in combination with his special implication $\rightarrow_L$ (see below). They are nowadays often called 'strong Kleene connectives'. Weak negation, to our knowledge, did not attract much attention until [Blau 1978] where it is motivated by natural language semantics. It is definable in many other systems of partial logic by means of certain other (rather 'exotic') operators.[2]

---

[2] See e.g. [Blamey 1986].

3.1.3 Some Remarks on Implication

Since there are two negations we can define two material implication operators. The version with strong negation leads to the implication operator used by Kleene [1952],

$$F \to_K G := \sim F \vee G$$

Yet, this operator is not reflexive (since $\sim F \vee F$ is not a tautology) and, moreover, does not satisfy the deduction theorem. At least the latter requirement can be considered as being fundamental for any 'real' implication. Therefore, e.g., Avron [1991] argues that this implication "is not an implication in any sense, and it is just one out of many connectives that are definable" from $\sim$ and $\vee$.

The second material implication,

$$F \supset G := -F \vee G$$

is the basis for the implication operator used by Lukasiewicz [1920] which is definable as $F \to_L G := (F \supset G) \wedge (\sim G \supset \sim F)$. Although $\supset$ satsifies the deduction theorem (see, e.g., [Avron 1991]) it does not behave like a real implication in other respects, namely, it is not persistent, as the following example demonstrates:

$$\{p(a), q(a), r(b)\} \models \forall x(p(x) \supset q(x))$$

i.e. every p is q, which does no longer hold if the information content of the model grows in the following way:

$$\{p(a), p(b), q(a), r(b)\} \not\models \forall x(p(x) \supset q(x))$$

It is a natural part of the meaning of implication that it is able to express subsumption between predicates. However, the extensional subsumption relationship expressed by $p(x) \supset q(x)$ may only hold at the moment and get lost by growing information not respecting it. Intensional subsumption, on the other hand, is invariant under information growth. That is, it corresponds rather to a persistent implication which in the framework of partiality has to be non-truthfunctional if it is required to satisfy the deduction theorem. The constructive logic **N** presented below (see 3.3) is equipped with such an implication.

Notice that $\langle \{0, u, 1\}, -, \sim, \wedge, \vee, \supset, \to_L, \rangle$ is precisely the three-element representative of *Quasi-Pseudo-Boolean algebras* defined in [Rasiowa 1974] as the algebraic models for 'constructive logic with strong negation'.

3.1.4 Natural Deduction

Natural deduction calculi seem to be best-suited to relate standard logics with KRSs. A natural deduction system defining valid inferences between a finite set of formulas and a formula can be viewed as a uniform KRS where

$$L_{\text{Repr}} = L_{\text{Query}} = L_{\text{Input}}$$

and

$$\mathbf{Upd}(X, F) = X, F := X \cup \{F\}$$

Such a system is given by a set of rules of the form

$$\frac{premise_1 \quad \ldots \quad premise_n}{conclusion}$$

where premises as well as the conclusion are derivation sequences of the form $X \vdash F$.

Derivability, then, can be defined by induction on the *length of a derivation*:

Definition 3.3 (Derivablility) *If $F \in X$ then F can be derived from X with length 0, symbolically $X \vdash^0 F$. F can be derived from X with length n, symbolically $X \vdash^n F$, if there is a rule with conclusion $X \vdash F$, and all derivations in the premise of this rule have a length $i < n$. A formula F, then, is said to be derivable from X, symbolically $X \vdash F$, if $X \vdash^n F$ for some $n < \omega$.*

The 4-valued logic $\mathbf{B}$ of Belnap [1977] with the basic connectives $\sim, \wedge$ and $\vee$ is given by the structural rules of Reflexivity, Cut and Weakening,

(Reflexivity) $F \vdash F$

(Cut) $\dfrac{X \vdash F \quad X, F \vdash G}{X \vdash G}$

(Weakening) $\dfrac{X \vdash G}{X, F \vdash G}$

and the following introduction and elimination rules:

$(\wedge)$ $\dfrac{X \vdash F \quad X \vdash G}{X \vdash F \wedge G}$ $\dfrac{X, F, G \vdash H}{X, F \wedge G \vdash H}$

$(\sim\wedge)$ $\dfrac{X \vdash \sim F}{X \vdash \sim(F \wedge G)} \quad \dfrac{X \vdash \sim G}{X \vdash \sim(F \wedge G)}$ $\dfrac{X, \sim F \vdash H \quad X, \sim G \vdash H}{X, \sim(F \wedge G) \vdash H}$

$(\sim\sim)$ $\dfrac{X \vdash F}{X \vdash \sim\sim F}$ $\dfrac{X, \sim\sim F \vdash G}{X, F \vdash G}$

The rules for disjunction are derivable according to the DeMorgan laws. The derivability relation $\vdash_B$ is the smallest relation between sets of formulas and formulas such that it is closed under the above rules. It is an adequate system with respect to 4-valued model-theoretic consequence, i.e. for $X \subseteq L(\sim, \wedge, \vee)$ and $F \in L(\sim, \wedge, \vee)$ we have

Claim 3.1 (Adequacy) $X \vdash_B F$ *iff* $X \models_4 F$

For a Henkin-style completeness proof see [Thijse 1992], where another proof
system is used, however.

3.1.5   Disjunctive Normal Form

Formulas of partial logic can be normalized in the same manner as those of
classical logic. For this purpose, DNS(F), the *disjunctive normal set* of a formula
F, is defined as follows:

$$
\begin{aligned}
\mathrm{DNS}(1) &= \{\emptyset\} \\
\mathrm{DNS}(e) &= \{\{e\}\} \\
\mathrm{DNS}(F \wedge G) &= \{E \cup D : E \in \mathrm{DNS}(F),\, D \in \mathrm{DNS}(G)\} \\
\mathrm{DNS}(F \vee G) &= \mathrm{DNS}(F) \cup \mathrm{DNS}(G)
\end{aligned}
$$

This formulation is inspired by a similar one (without weak negation) in [Miller
1989]. As explained above, all other cases of compound formulas can be handled by the resp. rewrite rules. The disjunctive normal form of a formula
$G \in L(1, -, \sim, \wedge, \vee)$ is obtained as

$$
\mathrm{DNF}(G) = \bigvee_{E \in \mathrm{DNS}(G)} \bigwedge E
$$

The disjunctive normalization steps are illustrated by an example:

Example 3.1

$$
\begin{aligned}
\mathrm{DNS}(p \wedge -(\sim q \wedge -r)) &= \{\{p\} \cup L : L \in \mathrm{DNS}(-(\sim q \wedge -r))\} \\
&= \{\{p\} \cup L : L \in \{\{-\sim q\}, \{r\}\}\} \\
&= \{\{p, -\sim q\}, \{p, r\}\}
\end{aligned}
$$

So, $\mathrm{DNF}(p \wedge -(\sim q \wedge -r)) = (p \wedge -\sim q) \vee (p \wedge r)$.

The disjunctive normal form of G is logically equivalent to G in the following
sense:

Claim 3.2 $\mathcal{M} \models \mathrm{DNF}(G)$ *iff* $\mathcal{M} \models G$

Proof: by induction on the complexity of G; I only show one case,

$$
\begin{aligned}
\mathcal{M} \models -\sim(F \wedge G) \quad &\text{iff} \quad \mathcal{M} \models -\sim F \text{ and } \mathcal{M} \models -\sim G \\
&\text{iff} \quad \mathcal{M} \models \mathrm{DNF}(-\sim F) \text{ and } \mathcal{M} \models \mathrm{DNF}(-\sim G)
\end{aligned}
$$

by the induction hypothesis

$$
\text{iff} \quad \mathcal{M} \models \bigvee_{E \in \mathrm{DNS}(-\sim F)} \bigwedge E \wedge \bigvee_{D \in \mathrm{DNS}(-\sim G)} \bigwedge D
$$

$$
\text{iff} \quad \mathcal{M} \models \bigvee_{\substack{E \in \mathrm{DNS}(-\sim F) \\ D \in \mathrm{DNS}(-\sim G)}} \bigwedge (E \cup D)
$$

$$
\text{iff} \quad \mathcal{M} \models \mathrm{DNF}(-\sim(F \wedge G)) \quad \square
$$

If E is a set of extended literals, E^+ consists of all members of E which are not weakly negated, $E^+ := E \cap \mathrm{Lit}$, and E^- collects the remaining weakly-negated literals, $E^- := \{l : -l \in E\}$. The following observation is an obvious consequence of the above normal form theorem:

Observation 3.6 $\mathcal{M} \models F$ *iff there is some* $E \in \mathrm{DNS}(F)$ *such that* $E^+ \subseteq M$ & $E^- \cap M = \emptyset$.

Notice that the DNS operator collects exactly the minimal models of a formula without weak negation, i.e. $\mathrm{MinMod}(F) = \mathrm{DNS}(F)$ for all $F \in L(1, \sim, \wedge, \vee)$. Consequently, for $X \subseteq L(1, \sim, \wedge, \vee)$ and $F \in L(1, -, \sim, \wedge, \vee)$,

$$
X \models_4 F \text{ iff } \forall L \in \mathrm{DNS}(\textstyle\bigwedge X) \, \exists E \in \mathrm{DNS}(F) : E^+ \subseteq L \ \& \ E^- \cap L = \emptyset
$$

Definition 3.4 *A ground formula* F *is called* definite *if its disjunctive normal set is a singleton, i.e. if* $\mathrm{DNS}(F) = \{E\}$ *for some* $E \subseteq \mathrm{XLit}$. *The set of all definite formulas of a given language* L *is denoted by* DefL.

Since there are two negations, various forms of inconsistency can be distinguished: *weak consistency* excludes weak complements like $p \wedge -p$, whereas *strong consistency* in addition excludes strong complements like $p \wedge \sim p$ which are considered to be contradictory.

Definition 3.5 *A ground formula* F *is called* weakly consistent *if for some* $E \in \mathrm{DNS}(F)$, $E \cap \overline{E} = \emptyset$.

Definition 3.6 *A ground formula* F *is called* consistent *if for some* $E \in \mathrm{DNS}(F)$, $E \cap \overline{E} = \emptyset$ *and* $E^+ \cap \widetilde{E^+} = \emptyset$.

3.1.6 Partial Logic and Classical Logic

Some Remarks on Computational Efficiency

Disappointingly, for arbitrary formulas $F, G \in L(\sim, \wedge, \vee)$, 4-valued entailment $\models_4$ is not easier to compute than its 2-valued counterpart $\models_2$. As noted in

[Patel-Schneider 1990], the problem of determining whether $F \models_4 G$ is co-NP complete, since the problem $F \models_2 G$ can be linearly reduced to it. However, the picture changes if F and G are in disjunctive normal form. Then computing whether

1. $F \models_4 G$ takes time proportional to the product of sizes of F and G, while

2. $F \models_2 G$ is still co-NP complete.

Apparently, there are two sources of intractability in the entailment problem: formula structure and reasoning structure. While the complexity of formula structure can be reduced by normalization, the complexity of reasoning can be reduced by using a sufficiently weaker logic (e.g. **B**), in other words, by dropping such computationally problematic rules of inference as *Tertium Non Datur* (resp. *Reasoning by Cases*, and *Reductio ad Absurdum*.

Tautologies

A tautology is a formula F which is valid in any model $\mathcal{M}$, $\mathcal{M} \models F$. An antitautology is a formula F which is invalid in any model $\mathcal{M}$, $\mathcal{M} \not\models F$.

In partial logic, the following are tautologies: $-(F \wedge -F)$, and $F \vee -F$, whereas neither $F \vee \sim F$ nor $\sim (F \wedge \sim F)$ is one. $\sim F \wedge \sim -F$ is an anti-tautology, but neither of $F \wedge -F$ nor $F \wedge \sim F$ is one.[3]

Literal Consequences

Proper partial (or 3-valued) logic and classical logic agree on literal consequences as Tan [1992] has pointed out:

Observation 3.7   *Let X be a finite set of formulas without weak negation, $X \subseteq L(\sim, \wedge, \vee)$, then $X \models_2 l$ iff $X \models_3 l$.*

Proof: ($\Leftarrow$) holds because classical consequence is charactzerized by 2-valued, i.e. total, models which are special cases of proper partial models. So, if every proper partial model that verifies X also verifies l, the same holds for all total models.

($\Rightarrow$) By the normal form theorem of classical logic, we obtain that $X \models_2 l$ iff $L \models_2 l$ for all $L \in \text{DNS}(\bigwedge X)$. Since the same holds for $\models_3$, it suffices to show that for arbitrary sets of literals L, $L \models_2 l$ implies $L \models_3 l$. If L is inconsistent this is trivially the case. Otherwise, $l \in M$ for all $M \in \text{Mod}_2(L)$, implying that $l \in \bigcap \text{Mod}_2(L) = L$. Since $L \subseteq M$ for all $M \in \text{Mod}_3(L)$, l is also valid in all proper partial models of L. $\square$

---

[3] Since there is no implication in the object language, implicative relations between formulas cannot be expressed by respective tautologies. They are, however, reflected in the deduction rules presented below.

This agreement of partial and classical logic for literal consequences is lost in the case of general partial logic: $K \models_2 l$ does not imply that $K \models_4 l$, simply because from an inconsistent K one does not get trivial consequences since it has a general partial model.

Translation into Classical Logic

A formula $F \in L(-, \sim, \wedge, \vee)$ can be translated into a corresponding formula $F^* \in L(\neg, \wedge, \vee)$ in a language with an extended set of predicate symbols in two steps:

1. Transform F into disjunctive normal form, $\text{DNF}(F)$.[4]

2. Obtain F^* from $\text{DNF}(F)$ by replacing

    (a) all literal subformulas $\sim p(t)$ with $p'(t)$ where p' is a newly introduced predicate symbol,

    (b) all occurences of $-$ with $\neg$.

In [Fenstad et al. 1987] this transformation is called *Feferman Translation Procedure*, and it is shown that the following correspondence holds:

$$X \models_4 F \;\Rightarrow\; X^* \models_2 F^*$$

where $X^* = \{F^* : F \in X\}$ and the negation sign $\neg$ in F^* denotes classical negation.

3.2   Belnap's Concept of an Information State

A KRS, according to Belnap, answers queries by invoking some inference mechanism, and it accepts input from a variety of sources by using an appropriate update mechanism. In such circumstances inconsistency threatens. Mr. X tells the KRS that A while Mrs. Y tells it that $\sim A$. Or, in a different environment, an automatic measurement yields that $m > 0.3$ while the subsequent confirmation attempt yields that $m < 0.3$.

What is the KRS to do ?

Possibility 1: Refuse to accept inconsistent information. However: this is unfair either to Mr. X or to Mrs. Y. Also, contradictions may not lie on the surface.

---

[4] In [Fenstad et al. 1987] negation normal form is used instead of disjunctive normal form. But this difference is not essential.

Possibility 2: Revise current beliefs in the presence of contradictions. However: it seems to be difficult to determine the proper revision policy doing justice to both Mr. X and Mrs. Y, and it seems to be even more difficult to mechanize it in a satisfactory way.

Possibility 3: Just accept contradictions and report them exactly as they were told, so the user can make up her mind.

Belnap advocates possibility 3, but emphasizes that even if the ultimate goal is possibility 2, i.e. revision, possibilty 3 is a good first step towards that goal.

3.2.1   Definite Epistemic States

Let X denote an epistemic state representing the current knowledge. If an atom a comes in as 'told True' the resulting expansion of X is simply

$$\mathbf{Upd}(X, a \text{ told True}) = X \cup \{a\}.$$

If a comes in as 'told False' then,

$$\mathbf{Upd}(X, a \text{ told False}) = X \cup \{\sim a\}.$$

In this way, by accepting only atomic input, an epistemic state X consisting of literals is obtained. It can be identified with a partial interpretation (negation at this level is a mere marker for disbelief).

Answers, then, are four-valued indeed,

$$\mathbf{Ans}(X, F) = \left\{ \begin{array}{llll} \text{yes} & \text{if} & X \models F & \& \ X \not\dashv F \\ \text{no} & \text{if} & X \dashv F & \& \ X \not\models F \\ \text{unknown} & \text{if} & X \not\dashv F & \& \ X \not\models F \\ \text{yes and no} & \text{if} & X \models F & \& \ X \dashv F \end{array} \right.$$

It should be clear that the answer 'yes and no' does not have the ontological force "That's the way the world is," but rather the epistemic force, "That's what I've been told."

Observation 3.8   *If input is restricted to atomic information (more precisely, literals) Belnap's concept of an epistemic state is a cumulative basic VKRS $\langle 2^{\text{Lit}}, \models, L(\sim, \wedge, \vee)\rangle$, i.e. the associated inference relation $\models$ is constructive and non-explosive, and only definite information is represented.*

Two principles of inference, both of which are essential in classical logic (but only the latter one in intuitionistic logic), fail in this setting:

(i) From an arbitrary sentence F one may conclude that for any sentence G, $G \vee \sim G$ holds.

(ii) From a pair of contradictory sentences, $F \wedge \sim F$, one may conclude any
   sentence G.

The failure of the first is needed if the inference relation is to be constructive.
The failure of the second is needed if the KRS is not to break down in the
presence of inconsistent information. Belnap calls these principles 'paradoxes
of implication'. Following the idea of relevance he has to reject them because
they lay the ground for non-relevant inferences (neither is F relevant for the
derivation of $G \vee \sim G$, nor is $F \wedge \sim F$ relevant for the derivation of G; in both
cases the principle of sharing is violated: neither of the two pairs of formulas
has anything in common between premise and conclusion).

Belnap goes on to define the update operation for arbitrary compound sentences which requires a more general notion of epistemic states. In order to
acommodate the representation of disjunctive information a set of partial interpretations is needed. Although vivid reasoning is primarily concerned with
definite information I will first present Belnap's general concept of an epistemic
state and then continue with his considerations on non-truthfunctional knowledge.

3.2.2   General Epistemic States

In order to process arbitrary input formulas $F \in L(\sim, \wedge, \vee)$, an epistemic state
Y has to consist of sets of literals (i.e. partial interpretations), $Y \subseteq 2^{Lit}$, representing the different epistemic alternatives. Following Belnap, yet formulated
differently, we get the following input processing:

$$\begin{aligned}
\mathbf{Upd}(Y, l) &= \{X \cup \{l\} : X \in Y\} \\
\mathbf{Upd}(Y, F \wedge G) &= \mathbf{Upd}(\mathbf{Upd}(Y, F), G) \\
\mathbf{Upd}(Y, F \vee G) &= \mathbf{Upd}(Y, F) \cup \mathbf{Upd}(Y, G)
\end{aligned}$$

All other cases of compound formulas are treated by rewriting.

Observation 3.9 $\mathbf{Upd}(Y, \mathrm{DNF}(F)) = \{X \cup K : X \in Y, K \in \mathrm{DNS}(F)\}$

Proof by simple rewriting:

$$\begin{aligned}
\mathbf{Upd}(Y, \mathrm{DNF}(F)) &= \mathbf{Upd}(Y, \bigvee_{K \in \mathrm{DNS}(F)} \bigwedge K) \\
&= \bigcup_{K \in \mathrm{DNS}(F)} \mathbf{Upd}(Y, \bigwedge K) \\
&= \bigcup_{K \in \mathrm{DNS}(F)} \{X \cup K : X \in Y\} \\
&= \{X \cup K : X \in Y, K \in \mathrm{DNS}(F)\} \quad \square
\end{aligned}$$

Upd respects disjunctive normal forms:

Claim 3.3 $\mathbf{Upd}(Y, F) = \mathbf{Upd}(Y, \mathrm{DNF}(F))$

Proof by induction on F:

For a literal l the assertion holds trivially.

For a conjunction $F \wedge G$ we get:

$\mathbf{Upd}(Y, F \wedge G)$
$= \mathbf{Upd}(\mathbf{Upd}(Y, F), G)$
$= \mathbf{Upd}(\mathbf{Upd}(Y, \mathrm{DNF}(F)), \mathrm{DNF}(G))$ by the induction hypothesis
$= \mathbf{Upd}(\{X \cup K : X \in Y, K \in \mathrm{DNS}(F)\}, \mathrm{DNF}(G))$ by observation 3.7
$= \{X \cup K \cup L : X \in Y, K \in \mathrm{DNS}(F), L \in \mathrm{DNS}(G)\}$
$= \mathbf{Upd}(Y, \displaystyle\bigvee_{\substack{K \in \mathrm{DNS}(F) \\ L \in \mathrm{DNS}(G)}} \bigwedge K \cup L)$

$= \mathbf{Upd}(Y, \mathrm{DNF}(F \wedge G))$

Finally, for a disjunction $F \vee G$,

$\mathbf{Upd}(Y, F \vee G)$
$= \mathbf{Upd}(Y, F) \cup \mathbf{Upd}(Y, G)$
$= \mathbf{Upd}(Y, \mathrm{DNF}(F)) \cup \mathbf{Upd}(Y, \mathrm{DNF}(G))$ by the induction hypothesis
$= \{X \cup K : X \in Y, K \in \mathrm{DNS}(F)\} \cup \{X' \cup L : X' \in Y, L \in \mathrm{DNS}(G)\}$
$= \{X \cup K : X \in Y, K \in \mathrm{DNS}(F) \cup \mathrm{DNS}(G)\}$
$= \mathbf{Upd}(Y, \mathrm{DNF}(F \vee G))$ □

As a consequence, the update operation can also be defined by means of the disjunctive normal set:

Observation 3.10 $\mathbf{Upd}(Y, F) = \{X \cup K : X \in Y, K \in \mathrm{DNS}(F)\}$

Hence, $\mathbf{Upd}$ satisfies Contraction and Permutation.

The next question concerns the answer operation, respectively the inference relation, associated with an epistemic state. Belnap defines that a general epistemic state takes a formula as something it has been told if it comes out true on each of the alternative interpretations it consists of:

Definition 3.7 $Y \vdash F \stackrel{def}{\Longleftrightarrow} X \models F$ *for all* $X \in Y$

Notice that $Y = \emptyset$ is not a meaningful epistemic state, it trivially confirms every sentence. The 'empty' epistemic state 0 is rather represented by $\{\emptyset\}$.

Observation 3.11 $C(Y_1 \cup Y_2) = C(Y_1) \cap C(Y_2)$

Epistemic states are not unique representations. Since $C(Y) = \bigcap\{C(\{X\}) : X \in Y\}$ and $C(\{X\}) \subseteq C(\{X'\})$ whenever $X \subseteq X'$, only the minimal elements of an epistemic state count.

Observation 3.12   *In general, $C(Y) = C(Y')$ does not imply that $Y = Y'$.*

Proof: For instance, $C(\{\{p\}\}) = C(\{\{p\}, \{p, q\}\})$. $\square$

Observation 3.13 $Y \vdash F$ *iff* $\forall X \in Y \, \exists K \in \text{DNS}(F) : K \subseteq X$

Definition 3.8 (Belnap's KRS)   *The KRS*

$$\langle 2^{2^{\text{Lit}}}, \vdash, L(\sim, \wedge, \vee), \textbf{Upd}, L(\sim, \wedge, \vee) \rangle$$

is called Belnap's KRS.

Claim 3.4   *Belnap's KRS satisfies Monotonicity, i.e. $C(Y) \subseteq C(\textbf{Upd}(Y, F))$.*

Proof by induction on F: For $F = l$ the assertion follows immediately from the persistence property (observation 3.2). For $F = G \wedge H$ we obtain by the induction hypothesis that

$$\begin{aligned}
C(Y) &\subseteq C(\textbf{Upd}(Y, G)) \\
&\subseteq C(\textbf{Upd}(\textbf{Upd}(Y, G), H)) \\
&= C(\textbf{Upd}(Y, G \wedge H))
\end{aligned}$$

Finally, for $F = G \vee H$ we get

$$\begin{aligned}
C(\textbf{Upd}(Y, G \vee H)) &= C(\textbf{Upd}(Y, G) \cup \textbf{Upd}(Y, H)) \\
&= C(\textbf{Upd}(Y, G)) \cap C(\textbf{Upd}(Y, H)) \\
&\supseteq C(Y)
\end{aligned}$$

since $C(Y) \subseteq C(\textbf{Upd}(Y, G))$ as well as $C(Y) \subseteq C(\textbf{Upd}(Y, H))$, by the induction hypothesis. $\square$

Claim 3.5   *Belnap's KRS is a cumulative basic VKRS.*

Proof: (Restricted Reflexivity) Reflexivity can be easily proved by induction on F.

(Cumulativity) Since Belnap's KRS is monotonic (see claim 3.4) it suffices to show Lemma Redundancy, i.e.

$$Y \vdash F \ \& \ \textbf{Upd}(Y, F) \vdash G \ \Rightarrow \ Y \vdash G$$

From observations 3.11 and 3.8 it follows that $Y \vdash F \ \Rightarrow \ Y \subseteq \textbf{Upd}(Y, F)$. Consequently, if $X' \models G$ for all $X' \in \textbf{Upd}(Y, F)$, then also $X \models G$ for all $X \in Y$, i.e. $Y \vdash G$.

(Constructivity) For constructible truth it has to be shown that for any $K \subseteq \text{Lit}$, if $F \vee G$ is derivable from $\lceil K \rceil = \{K\}$, then F or G is derivable from it. Since $\{K\} \vdash F$ iff $K \models F$ this follows from the definition of $\models$. Similarly in the case of constructible falsity.

(Non-Explosiveness) It has to be shown that for any non-tautology F and any epistemic state Y there is a variant F' of F such that $Y \not\vdash F'$. Since $X \models F$ only if F and X have symbols in common it suffices to substitute in F all constituents sharing symbols with Y by symbols not occuring in Y. The resulting variant of F, then, cannot be derivable from Y. $\square$

In 4.6 weak negation will be added to Belnap's KRS in a straightforward manner. However, although a well-defined inference relation is obtained, Lemma Compatibility no longer holds.

The process of information growth can be captured by the following notion of informational extension. An epistemic state Y' is called an *(informational) extension* of Y, symbolically $Y' \geq Y$, if every model in Y' extends some model in Y:

Definition 3.9 $\quad Y' \geq Y \overset{def}{\Longleftrightarrow} \forall X' \in Y' \exists X \in Y : X' \geq X$

Observation 3.14 $\quad \langle 2^{2^{\mathrm{Lit}}}, \geq \rangle$ *is a preordering (i.e. reflexive and transitive relation) with least element* $0 := \{\emptyset\}$.

Observation 3.15 $\quad \vdash$ *is persistent, i.e. if* $Y' \geq Y$ *then* $Y' \vdash F$ *whenever* $Y \vdash F$ *for all* $F \in L(\sim, \wedge, \vee)$.

When an update takes place, the current information never decreases, in other words, the update operation is *ampliative*:

Observation 3.16 $\quad Y \leq \mathbf{Upd}(Y, F)$

In order to check entailment between formulas,[5] it suffices to check the derivability of G from the empty KB updated by F because $\vdash$ is monotonic:

Observation 3.17 $\quad F \vdash G \quad$ *iff* $\quad \lceil F \rceil \vdash G$

Entailment between formulas corresponds to a greater information content in the following sense:

Observation 3.18 $\quad F \vdash G \quad$ *iff* $\quad \lceil F \rceil \geq \lceil G \rceil$

which follows immediately from

Observation 3.19 $\quad \lceil F \rceil = \mathrm{DNS}(F)$

---

[5] $F \vdash G$ denotes entailment between formulas induced by the inference relation $\vdash \subseteq L_{\mathrm{KB}} \times L(\sim, \wedge, \vee)$ as described in observation 2.5.

Proof: By an easy calculation,

$$
\begin{aligned}
\lceil F \rceil \; &= \; \mathbf{Upd}(0, F) \\
&= \; \bigcup_{K \in \mathrm{DNS}(F)} \{X \cup K : X \in 0\} \\
&= \; \bigcup_{K \in \mathrm{DNS}(F)} \{K\} \\
&= \; \mathrm{DNS}(F) \quad \square
\end{aligned}
$$

3.2.3   Information States

Belnap gives a further generalization of the concept of an epistemic state, called *information state*, where he also considers conditional information in addition to the extensional information represented by an epistemic state. He defines a rule in a very general way, namely as 'any continous and ampliative mapping from epistemic states into epistemic states'. Recall that a subset of a lattice is called *directed* if every pair of elements has a common upper bound in it. A mapping from a lattice into another one, $f : A \to B$, is *continous* if for all nonempty directed $X \subseteq A$, $f(\sup X) = \sup\{f(x) : x \in X\}$. If A is finite this is equivalent to the simpler requirement that f be *monotonic*, $x \leq y \;\Rightarrow\; f(x) \leq f(y)$. A mapping from a lattice into itself, $f : A \to A$ is *ampliative* if for all x, $x \leq f(x)$.

Provided that epistemic states are finite (on the basis of a finite Herbrand universe), the notion of a continous rule can be replaced with that of a monotonic one for our purposes. Intuitively, an ampliative rule r never decreases the amount of information represented by an epistemic state, $Y \leq r(Y)$, while a monotonic rule respects any increase of information.

An *information state*, according to Belnap, is a pair $\langle Y, R \rangle$ consisting of an epistemic state Y and a set of rules R. One can think of Y and R as the extensional and intensional knowledge represented by the information state. While the extensional knowledge is static (describing a certain epistemic state of the world), the intensional knowledge is dynamic: it provides for internal updates of the extensional knowledge in case of updates from the outside. Whenever Y is changed R can be invoked in order to 'saturate' the extensional knowledge according to the principle of 'minimal mutilation': only as much additional extensional knowledge is produced by R so as to satisfy all rules in R. A rule r is *satisfied* in Y, or in other words, Y is closed under r, if $r(Y) = Y$, i.e. if r does not contribute anything to Y.

How does R work in order to contribute to the extensional knowledge Y ? Let $\hat{R}$ be the closure of R under composition. $\hat{R}$ is a directed set, since the composition of two ampliative monotonic functions will always provide an upper bound for both of them. Therefore, the limit $\hat{r} := \sup \hat{R}$ is also an element of $\hat{R}$.

Claim 3.6 $\hat{r}(Y)$ *is the 'minimum mutilation' of Y in which all rules $r \in R$ are satisfied.*

Proof: It has to be shown that for any $r \in R$, $r(\hat{Y}) = \hat{Y}$ where $\hat{Y} = \hat{r}(Y)$. Since by definition $r \leq \hat{r}$, $\hat{Y} \subseteq r(\hat{Y}) \subseteq \hat{r}(\hat{Y}) = \hat{Y}$. $\square$

For the sake of conveniency, I write $R(Y)$ instead of $\hat{r}(Y)$. It is justified, then, to say that the epistemic state $R(Y)$ is the extensional meaning of $\langle Y, R \rangle$, and to define,

$$\mathbf{Ans}(\langle Y, R \rangle, F) := \mathbf{Ans}(R(Y), F).$$

Observation 3.20   *Two information states, $\mathbf{I}_1 = \langle Y_1, R_1 \rangle$ and $\mathbf{I}_2 = \langle Y_2, R_2 \rangle$, are answer equivalent iff they have the same closure,*

$$\mathbf{I}_1 \simeq \mathbf{I}_2 \quad iff \quad R_1(Y_1) = R_2(Y_2)$$

3.2.4   Rule-Based Systems as Information States

A rule-based system, or rule knowledge base, $\langle Y, R \rangle$, consisting of a set of (compiled) facts Y and a set of monotonic rules R, can be viewed as an information state in the sense of Belnap. Real world applications, however, require the use of nonmonotonic – and, in certain cases, non-ampliative – rules. In normal logic programs, for instance, where negation-as-failure is admitted in the premise of a rule and therefore premise formulas may be not persistent, Belnap's requirement that rules have to be continous mappings is violated.

Observation 3.21   *By admitting weak negation in the premise (and eventually in conclusions) of rules, all the standard regularities, such as monotonicity and ampliativeness, guaranteeing the existence of fixpoints according to Scotts theory of continous lattices get lost although Belnap's general framework of information states is retained. Other regularites which would allow to obtain the desired set of consequences as a fixpoint even in the presence of nonmonotonic and non-ampliative rules have to be identified.*

3.3   Paraconsistent Constructive Logic

Partial logic has the obvious lack of a genuine implication connective satisfying the deduction theorem and being persistent (see 3.1.3). The paraconsistent constructive logic $\mathbf{N}^6$ is the conservative extension of the partial logic $\mathbf{B}$ by the addition of intuitionistic implication. $\mathbf{B}$ and $\mathbf{N}$ can be considered as the base logics for information processing and vivid reasoning.

There are two independent lines of argumentation in favour of $\mathbf{N}$:

---

[6] Which is called N^- in [Almukdad & Nelson 1984].

1. **N** improves partial logic (more specifically, **B**) by extending it by a genuine implication connective yielding greater expressivity and preserving the general information processing perspective (notably constructivity and non-explosiveness).

2. **N** also improves *positive logic*, a subsystem of intuitionistic logic which is important for standard databases and logic programs where only positive information is represented, by adding to it a negation connective yielding greater expressivity and allowing for the representation of explicit negative information.

Moreover, **N** is preferable over intuitionistic logic since its negation is truly constructive whilst intuitionistic negation is not,[7] which means that **N** is cognitively more fundamental and computationally more feasible than intuitionistic logic.

3.3.1 Partial Kripke Models for Constructive Logic

Since the semantics of intuitionistic implication is not truth-functional, models for constructive logic must be able to represent the dynamics of information growth, e.g. in the style of Kripke structures. A *constructive Kripke frame*, thus, is a pair $\langle Y, \preceq \rangle$ where Y is a set of partial worlds, $Y \subseteq 2^{\text{Lit}}$, and $\preceq$ is a binary relation on Y satisfying $M_1 \preceq M_2 \Rightarrow M_1 \subseteq M_2$.[8]

Models consist of a frame $\mathcal{K} = \langle Y, \preceq \rangle$ and a reference world $M \in Y$. They are defined here for ground formulas only. Atoms, conjunctions and negations are verified and falsified on the spot:

$$
\begin{aligned}
\mathcal{K}, M &\models a & &\text{iff} & a &\in M \\
\mathcal{K}, M &\models F \wedge G & &\text{iff} & \mathcal{K}, M &\models F \text{ and } \mathcal{K}, M \models G \\
\mathcal{K}, M &\models {\sim} F & &\text{iff} & \mathcal{K}, M &\dashv F \\
\mathcal{K}, M &\dashv a & &\text{iff} & {\sim} a &\in M \\
\mathcal{K}, M &\dashv F \wedge G & &\text{iff} & \mathcal{K}, M &\dashv F \text{ or } \mathcal{K}, M \dashv G \\
\mathcal{K}, M &\dashv {\sim} F & &\text{iff} & \mathcal{K}, M &\models F
\end{aligned}
$$

Implications are verified with respect to the possible growth of information according to

$$
\mathcal{K}, M \models F \to G \quad \overset{def}{\Longleftrightarrow} \quad \text{for all } M' \in Y \text{ such that } M' \geq M:
$$
$$
\mathcal{K}, M' \models F \Rightarrow \mathcal{K}, M' \models G
$$

---

[7] This is discussed, e.g., in [Pearce 1991] and [Wansing 1992].

[8] This is the partial version of the kind of Kripke structures that have been called 'research' interpretations by Grzegorczyk [1964] who argued that "intuitionistic logic can be understood as the logic of scientific research" where a "scientific research (e.g. an experimental investigation) consists of the successive enrichment of the set of data by new established facts obtained by means of our method of inquiry". Since Grzegorczyk considers the general case (where the language is infinite) he has to require that Y is a set of finite sets.

and falsified on the spot according to

$$\mathcal{K}, M \dashv F \to G \overset{def}{\Longleftrightarrow} \mathcal{K}, M \vDash F \ \& \ \mathcal{K}, M \dashv G$$

It is convenient to say that F holds in a frame $\mathcal{K} = \langle Y, \preceq \rangle$, symbolically $\mathcal{K} \vDash F$, if $\mathcal{K}, M \vDash F$ for all $M \in Y$.

The model-theoretic consequence $\vDash_N$ associated with this Kripke-style semantics is defined by

$$X \vDash_N F \quad \overset{def}{\Longleftrightarrow} \quad \text{for any constructive Kripke frame } \mathcal{K},$$
$$\text{if } \mathcal{K} \vDash G \text{ for all } G \in X, \text{then } \mathcal{K} \vDash F$$

Observation 3.22   *For extensional formulas, $F \in L(\sim, \wedge, \vee)$, the dynamic model relation coincides with the static one,*

$$\mathcal{K}, M \vDash F \quad iff \quad M \vDash F$$

3.3.2   Natural Deduction

In order to accommodate the natural deduction system $\vdash_B$ it has to be extended by positive and negative introduction and elimination rules for $\to$, resulting in $\vdash_N$:

$$(\to) \qquad \frac{X, F \vdash G}{X \vdash F \to G} \qquad\qquad \frac{X \vdash F \quad Y, G \vdash H}{X, Y, F \to G \vdash H}$$

$$(\sim\to) \qquad \frac{X \vdash F \quad X \vdash \sim G}{X \vdash \sim(F \to G)} \qquad\qquad \frac{X, F, \sim G \vdash H}{X, \sim(F \to G) \vdash H}$$

Since in **N** there is no semantical link between an atom and its negation – they are treated as independent pieces of information – inferences in **N** are essentially inferences in positive logic. The system $\vdash_P$ of positive logic is determined by the rules (Reflexivity), (Weakening), (Cut), ($\wedge$), ($\vee$), and ($\to$). It is well-known[9] that in **N** a formula $F \in L(\sim, \wedge, \vee, \to)$ can be transformed into an equivalent positive formula $F^* \in L(\wedge, \vee, \to)$ by introducing for each predicate symbol p a new predicate symbol p' which is used to represent the negation of p :

$$(p(t))^* \ = \ p(t)$$
$$(\sim p(t))^* \ = \ p'(t)$$

For compound formulas the negation symbol $\sim$ is driven in so that it stands immediately before atoms:

$$(F \circ G)^* \ = \ F^* \circ G^* \qquad \text{for } \circ = \wedge, \vee, \to$$

---

[9] See e.g. [Rautenberg 1979].

$$(\sim(F \wedge G))^* = (\sim F)^* \vee (\sim G)^*$$
$$(\sim(F \vee G))^* = (\sim F)^* \wedge (\sim G)^*$$
$$(\sim(F \rightarrow G))^* = F^* \wedge (\sim G)^*$$
$$(\sim\sim F)^* = F^*$$

Claim 3.7 *Let $F \in L(\sim, \wedge, \vee, \rightarrow) \supseteq X$. Then*

$$X \vdash_N F \Rightarrow X^* \vdash_P F^* \qquad where\ X^* = \{F^* : F \in X\}.$$

Proof: by straightforward induction on the length of a derivation.

This reducibility to positive logic is a very strong property which does not hold in most other logics where certain laws, notably (ECSQ), (Disjunctive Syllogism), (Contraposition), (Negation as Inconsistency), (Reductio ad Absurdum), and (Reasoning by Cases), stipulate semantical links between a sentence and its negation. This is illustrated by the following examples, where $\vdash_3$ stands for Kleene's 3-valued logic, $\vdash_J$ for Johannson's minimal logic, and $\vdash_{cl}$ for classical logic:

$$\{p, \sim p \vee q\} \vdash_3 q, \qquad \text{by (Disjunctive Syllogism)},$$
but $\qquad \{p, p' \vee q\} \not\vdash_P q.$

$$\{\sim q, p \rightarrow q\} \vdash_J \sim p, \qquad \text{by (Contraposition)},$$
but $\qquad \{q', p \rightarrow q\} \not\vdash_P p'.$

$$\{p \rightarrow q, \sim p \rightarrow q\} \vdash_{cl} q, \qquad \text{by (Reasoning by Cases)},$$
but $\qquad \{p \rightarrow q, p' \rightarrow q\} \not\vdash_P q.$

In [Pearce 1991] the adequacy of $\vdash_N$ with respect to $\models_N$ is shown (via translation to positive logic):

Claim 3.8 (Adequacy) *Let $F \in L(\sim, \wedge, \vee, \rightarrow) \supseteq X$, then*

$$X \vdash_N F \quad \textit{iff} \quad X \models_N F$$

3.3.3 Definite Information

It is well-known that the inference relation of constructive logic satisfies both the disjunction property and the property of constructible falsity (see e.g. [Rautenberg 1979]). It is also well-known that a KB consisting of Harrop formulas is definite in intuitionistic logic. The same holds for constructive logic.

Definition 3.10 (Harrop Formulas) *The set Hrp $\subseteq L(\sim, \wedge, \vee, \rightarrow)$ of Harrop formulas is the smallest set such that*

(H0) Lit $\subseteq$ Hrp.

(H1) If $F, G \in \mathrm{Hrp}$ then $\sim\sim F, F \wedge G, \sim(F \vee G) \in \mathrm{Hrp}$.

(H2) If $F \in L(\sim, \wedge, \vee, \rightarrow)$, and $G \in \mathrm{Hrp}$, then $F \rightarrow G \in \mathrm{Hrp}$.

(H3) If $F, \sim G \in \mathrm{Hrp}$ then $\sim(F \rightarrow G) \in \mathrm{Hrp}$.

Claim 3.9   *Let X be a set of Harrop formulas. Then*

$$X \vdash_N F \vee G \ \Rightarrow \ X \vdash_N F \ or \ X \vdash_N G$$

Proof: Can be obtained from the proof of a corresponding theorem for intuitionistic logic in [Prawitz 1965, p.55].

3.3.4   Disjunctive Normal Form

Since constructive implication is non-truthfunctional, normal forms in constructive logic cannot reduce formulas to their literal constituents as in partial logic. Nevertheless there is a corresponding normalization of arbitrary formulas. The irreducible elements of a normal form in constructive logic are *conditional facts*, i.e. expressions of the form $F \rightarrow l$ where F can be normalized as well if necessary.

$$
\begin{aligned}
\mathrm{DNS}(1) &= \{\emptyset\} \\
\mathrm{DNS}(l) &= \{\{1 \rightarrow l\}\} \\
\mathrm{DNS}(F \wedge G) &= \{K \cup L \ : \ K \in \mathrm{DNS}(F), \ L \in \mathrm{DNS}(G)\} \\
\mathrm{DNS}(\sim(F \wedge G)) &= \mathrm{DNS}(\sim F) \cup \mathrm{DNS}(\sim G) \\
\mathrm{DNS}(F \rightarrow G) &= \{\{\mathrm{DNF}(F \wedge H) \rightarrow l : H \rightarrow l \in K\} : K \in \mathrm{DNS}(G)\} \\
\mathrm{DNS}(\sim(F \rightarrow G)) &= \mathrm{DNS}(F \wedge \sim G) \\
\mathrm{DNS}(\sim\sim F) &= \mathrm{DNS}(F)
\end{aligned}
$$

The disjunctive normal form of a formula G is obtained as

$$\mathrm{DNF}(G) = \bigvee_{K \in \mathrm{DNS}(G)} \bigwedge K$$

which is conjectured to be logically equivalent to G in the following sense:

Conjecture 1 $\mathcal{K}, M \models \mathrm{DNF}(G)$ *iff* $\mathcal{K}, M \models G$

As in partial logic, a definite formula in constructive logic can also be characterized by its disjunctive normal set being a singleton:

Observation 3.23   *If F is a Harrop formula, then $\mathrm{DNS}(F)$ is a singleton.*

Example 3.2   *The Harrop formula $F = \sim(p \rightarrow q \vee r) \wedge (\sim q \rightarrow (r \rightarrow p \wedge \sim s))$ is definite:*

$$\mathrm{DNS}(F) = \{\{p, \sim q, \sim r, \sim q \wedge r \rightarrow p, \sim q \wedge r \rightarrow \sim s\}\}$$

Chapter 4

Vivid Reasoning on the Basis of Facts

A vivid knowledge representation system for the processing of definite extensional information, V_o, is presented as the base case of VKRSs. By adding a mechanism for the neutralization of contradictory information built into the update, resp. the inference, operation two further systems are obtained. They can be viewed as different possibilities of extending RDBs by allowing for explicit negative information with respect to inconsistency handling.

The distinction between the partial and total representation of predicates is introduced and used as a semantical basis for the Closed-World Assumption. It is shown by means of an example that the partial representation of an exact predicate requires disjunctive information, and leads to indefinite answers.

Finally, Belnap's KRS is 'vivified' by the addition of weak negation, yielding the disjunctive VKRS V_B. Remarkably, an epistemic state is a unique representation in V_B, which is not the case in Belnap's KRS. Also interesting, but maybe less satisfactory, is the fact that V_B does not satisfy Lemma Compatibility.

4.1 Fact Bases

Usually, facts are considered to be atomic sentences expressing the basic positive information to be represented. If this view of basic (irreducible) information is generalized to the fundamental dichotomy of two kinds of basic pieces of information, namely positive and negative ones, it is natural to speak of positive and negative facts expressed by the resp. literals. So, the more general notion of a fact adopted here is that of a literal sentence.

A KB consisting of positive and negative facts is actually a slight generalization of a *vivid knowledge base* in the restricted sense of [Levesque 1988], where only positive facts, i.e. ground atoms, are allowed.

Example 4.1 $X_1 = \{\sim b(S), m(P, L)\}$ *represents the information that Susan is not blonde, and that Peter is married to Linda.*

Definition 4.1 (Informational Ordering)   *Let X and X' be fact bases, i.e. $X, X' \subseteq \text{Lit}$. Then X' is an informational extension of X, symbolically $X' \geq X$, if $X' \supseteq X$.*

As a kind of natural deduction from positive and negative facts a derivability relation $\vdash$ between a set of proper ground literals X and a ground formula $F \in L(1, -, \sim, \wedge, \vee)$ is defined. Recall that it is only needed to give a definition for extended literals which can be extended to the case of general query formulas as described in 2.10.

Definition 4.2   *Derivability of a literal, resp. weakly negated literal, is defined by membership, resp. non-membership:*

$$
\begin{array}{llll}
(l) & X \vdash l & \text{if} & l \in X \\
(-l) & X \vdash -l & \text{if} & l \notin X
\end{array}
$$

For example, $X_1 \vdash \sim b(S) \wedge -\sim b(L)$. Notice that according to this definition both a literal l and its complement $\bar{l}$ are acceptable at the same time, and independently from each other. This kind of inference is called *liberal*, as opposed to certain neutralization-based inference procedures (see e.g. [Wagner 1991c]) where contradictory information is discarded.

Observation 4.1   *A liberal inference relation is not coherent, i.e. $X \vdash \sim p$ does not imply that $X \vdash -p$. In other words, it is possible to infer contradictory queries, such as $p \wedge \sim p$.*

Definition 4.3   *The update operation for extended literals is defined as insertion, resp. deletion:*

$$
\begin{array}{lll}
\mathbf{Upd}(X, l) & := & X \cup \{l\} \\
\mathbf{Upd}(X, -l) & := & X - \{l\}
\end{array}
$$

Notice that an update by a weakly negated fact leads to a deletion. This may look strange to a logician. In databases, however, deletion is as natural an update operation as insertion.

More generally, an input formula can be any definite formula $F \in DefL(1, -, \sim, \wedge, \vee)$, such that $\text{DNS}(F) = \{E_F\}$. Then, update is defined as

$$
\mathbf{Upd}(X, F) := X \cup E_F^+ - E_F^-
$$

For example, if we learn that Peter gets divorced from Linda and marries Susan, we perform the following update:

$$\mathbf{Upd}(X_1, -m(P,L) \wedge m(P,S)) = \{\sim b(S), m(P,S)\}$$

Such a sequence of basic inputs (insertions and deletions of atoms or literals) is called a *knowledge base transaction*.

Since $X \vdash F$ implies that $E_F^+ \subseteq X$ and $E_F^- \cap X = \emptyset$, we get

Observation 4.2 $X \vdash F \Rightarrow \mathbf{Upd}(X,F) = X$

Definition 4.4   *The KRS* $\langle 2^{\mathrm{Lit}}, \vdash, L(1,-,\sim,\wedge,\vee), \mathbf{Upd}, \mathrm{XLit} \rangle$ *is denoted by* V_o^-.

Claim 4.1 V_o^- *is a cumulative VKRS.*

Proof: Simple inspection of the inductive cases shows that V_o^- is reflexive for weakly consistent definite formulas. Observation 4.2 implies that V_o^- is cumulative. It is also constructive since both disjunctions and negated conjunctions can only be established by the rule $(\vee)$. It is non-explosive since every derivable formula can be made non-derivable by substituting symbols occuring in the KB with symbols not occuring in KB. $\square$

Observation 4.3   *In* V_o^-, $\mathbf{Upd}$ *does not satisfy Permutation, as a simple example shows:*

$$\mathbf{Upd}(\mathbf{Upd}(\emptyset, -p), p) = \{p\} \neq \mathbf{Upd}(\mathbf{Upd}(\emptyset, p), -p) = \emptyset$$

In order to have regular conjunction processing, therefore, Conjunction Composition cannot be used as a definition. Commutativity of conjunctive input is enforced instead by means of normalization. The net effect of this is that two inputs at different time points cannot be captured by conjunction, in general.

In a sense, a fact base X represents its own adequate model $\mathcal{M}_X = \langle X^t, X^f \rangle$ by setting $X^t = \{a : a \in X\}$ and $X^f = \{a : \sim a \in X\}$. Thus, we immediately obtain the following adequacy result:

Observation 4.4 $X \vdash F$ *iff* $\mathcal{M}_X \models F$.

which can be proved by straightforward induction on the complexity of F.

4.2   Inherent Consistency in Fact Bases

While in liberal reasoning both a query and its negation may be inferred simultaneously, an inherently consistent KRS based, e.g., on a *conservative inference*

procedure or a *consistency preserving update* operation blocks contradictory conclusions by neutralizing inconsistent information. A literal l can only be inferred
conservatively if its complement $\tilde{l}$ can not.[1] Both approaches capture the idea
that neither the 'old' (already represented) nor the 'new' (incoming) information
is to be preferred. Both systems are coherent, i.e.

$$X \vdash \sim p \;\Rightarrow\; X \vdash -p \;\Rightarrow\; X \not\vdash p$$

In other words, it is not possible that $X \vdash p \wedge \sim p$.

Definition 4.5   *For $L \subseteq$ Lit we define: $Cons(L) := L - \tilde{L}$ collecting the
consistent literals in L.*

Definition 4.6 (Consistency in a Fact Base)   *A formula F is* consistent
in a fact base X if there is $E \in \mathrm{DNS}(F)$ such that

1. *$E \cap \overline{E} = \emptyset$, and*

2. *$E^+ \cap \widetilde{E^+} = \emptyset$, and*

3. *$X \cap \widetilde{E^+} = \emptyset$.*

4.2.1 Deleting Contradictory Information

If contradictory information is considered to be no information the KB should
not contain contradictory literals since they are useless, and consequently, KB $\in$
$Cons(2^{\mathrm{Lit}})$ where the latter set denotes the set of all consistent sets of literals.
The non-derivability of contradictions is 'automatically' achieved by not allowing
the KB to become inconsistent after updates. Therefore, literals in the KB have
to be deleted if they are inconsistent with incoming information:

$$X \oplus l \;\; := \;\; \begin{cases} X \cup \{l\} & \text{if } \tilde{l} \notin X \\ X - \{\tilde{l}\} & \text{otherwise} \end{cases}$$

Updating with weakly negated literals, as well as the derivability of literals and
weakly negated literals, are defined in the same way as in the liberal system (i.e.
as deletion, membership test and non-membership test). The KRS

$$\langle\, Cons(2^{\mathrm{Lit}}), \vdash, L(-, \sim, \wedge, \vee), \oplus, \mathrm{XLit} \,\rangle$$

is denoted by $\boldsymbol{V}_{o,\oplus}^{-}$.

Observation 4.5 $X \vdash F$ *implies that*

---

[1] In general, one needs a two-level architecture based on the notions of support and ac-
ceptance, as decribed in [Wagner 1991c], where a literal is accepted if it is supported and its
complement is not.

1. F is consistent in X, and

2. $\exists K \in \mathrm{DNS}(F) : K^+ \subseteq X$ & $K^- \cap X = \emptyset$.

Claim 4.2 $V^-_{0,\oplus}$ *is a cumulative VKRS.*

Proof: See proof of claim 4.1. As for V^-_0, Cumulativity follows immediately from the fact that $X \vdash F$ implies $X \oplus F = \mathbf{Upd}(X, F) = X$. $\square$

4.2.2 Keeping Contradictory Information

It is also possible to keep contradictory information in the KB considering it to be overdetermined and excluding it from derivations. This is achieved by an inherently consistent inference procedure, called *conservative*, which does not accept contradictory information in the derivation of a conclusion but accepts it as conclusive evidence for its failure:

$$
\begin{array}{llll}
(l) & X \vdash_c l & \text{if} & l \in X \text{ and } \tilde{l} \notin X \\
(-l) & X \vdash_c -l & \text{if} & l \notin X \text{ or } \tilde{l} \in X
\end{array}
$$

It is not clear, however, how an update operation which is suitable for $\vdash_c$ has to be defined. As a first attempt, let $\mathbf{Upd}$ be as in the liberal system V^-_0 above, that is, updates would be simple insertions and deletions of literals, no matter what their epistemic status is in the KB. Call this system $V_{0,c}$.

Observation 4.6   *In $V_{0,c}$, lemmas are not redundant: $\{p, \sim p\} \vdash_c -p$, but*

$$
\mathbf{Upd}(\{p, \sim p\}, -p) = \{\sim p\} \vdash_c \sim p
$$

whereas $\{p, \sim p\} \nvdash_c \sim p$.

This violation of Lemma Redundancy seems to be rather a bug than a feature. It seems to be caused by the violation of another principle, namely that of *minimal change* which governs all belief change operations: the deletion of p from $\{p, \sim p\}$ is not necessary in order to make $-p$ derivable. On the other hand, it seems desirable to have the possibility to change the epistemic status of p from being overdetermined in a KB to being accepted or rejected, i.e. there should be an update of $\{p, \sim p\}$ yielding $\{p\}$. The further discussion of these questions will be an issue of future work.

Observation 4.7 $V_{0,c}$ *is not equivalent to $V_{0,\oplus}$: in the latter it holds that*

$$
((0 \oplus p) \oplus \sim p) \oplus p = \{p\} \vdash p
$$

while in in the former it holds that

$$
\mathbf{Upd}(\mathbf{Upd}(\mathbf{Upd}(0, p), \sim p), p) = \{p, \sim p\} \nvdash_c p
$$

4.3   Partial and Total Representation of Predicates

Körner [1966] has developed a notion of *inexact predicate* in order to account for the fact that with respect to empirical predicates the classical principle *tertium non datur* is not adequate. In the general case, then, one cannot conclude from the fact that a sentence is not true that it is therefore false. Inexact predicates have truth value gaps, that is, for a certain individual c it may be the case that neither $\sim p(c)$ nor $p(c)$ holds. For exact predicates, however, we have exactly one of $\sim p(c)$ or $p(c)$. In [Cleave 1974] the notion of logical consequence in the 3-valued logic of inexact predicates is discussed, and in [Almukdad & Nelson 1984] the close connection between Cleave's system and Nelson's constructive logic is mentioned.

Instead of 'inexact' and 'exact' I will also speak of *partial* and *total* predicates, thereby emphasizing a conceptual rather than ontological level. In order to handle total predicates by means of the Closed-World Assumption, a fact base has to define which predicates are positive-totally and which ones are negative-totally represented, e.g. by listing them in two resp. sets of predicate symbols as in the formulation of the CWA in 2.10.

A vivid fact base, then, is a triple $\langle \text{CWA}, X, IC \rangle$, consisting of a pair of predicate symbol sets, a set of facts and a set of integrity constraints such that for an interpretation $\mathcal{M}$, in order to be a model,

$$\mathcal{M} \models \langle \text{CWA}, X, IC \rangle$$

it must hold that

1. $X \subseteq M$, and

2. for all $p \in \text{CWA}^+$, and for all $t \in U_X$, if $X \vdash -p(t)$ then $\sim p(t) \in M$, and

3. for all $q \in \text{CWA}^-$, and for all $t \in U_X$, if $X \vdash -\sim q(t)$ then $q(t) \in M$, and

4. for all $F \in IC$, $\mathcal{M} \models F$.

where U_X denotes the Herbrand universe of X, i.e. the set of all individual constants occuring in X.

While many empirical predicates are inexact, i.e. necessarily partial in any possible representation, analytical predicates[2] are exact, and hence, can be totally represented. The above rule says that if we deal with a total representation of an exact predicate we can apply the CWA in order to check falsity or truth, respectively.

---

[2] E.g. legally defined predicates, like *married*.

4.4 Example

The following example demonstrates that already in the RDB-like setting of fact bases the distinction between explicit and implicit negative information makes sense and is quite useful.

$$X \quad = \quad \left\{ \begin{array}{l} \text{CWA} = \{married\} \\ IC = \{-\exists x : \text{married}(x, x)\} \\ \text{married(Peter, Mary)} \\ \text{blonde(Susan)} \\ \sim\text{blonde(Peter)} \end{array} \right.$$

X is the fact base used by the state authorities of a small country, where only one couple is registered as married, and it is stipulated by law that only registered marriages are legally valid, i.e. $married \in \text{CWA}^+$. The same state authorities did not stipulate yet who is blonde and who is not. They left this question open for case-by-case decisions based on empirical information, i.e. $blonde \notin \text{CWA}^+$. When they learn, however, that someone is blonde, or that she is definitely not, they eagerly store this information in their fact base. So, it is not clear to these authorities, whether it is false that Mary is blonde, only

$$X \vdash -\text{blonde(Mary)}$$

holds, but not

$$X \vdash \sim\text{blonde(Mary)}$$

whereas there is no question whether Peter and Susan are married,

$$X \vdash \sim\text{married(Peter,Susan)}$$

although this is not explicitly represented in X, but can be inferred from it by the CWA rule.

The least model of X (modulo irreflexivity and symmetry of $married$) is

$$M \quad = \quad \left\{ \begin{array}{l} \text{married(Peter,Mary)} \\ \text{blonde(Susan)} \\ \sim\text{blonde(Peter)} \\ \sim\text{married(Peter,Susan)} \end{array} \right.$$

4.5 Exact Predicates and Indefinite Answers

I will now discuss the phenomena of indefinite answers in connection with exact predicates by a reinterpretation of an example of Moore [1982]. Let

$$X = \{\text{on}(a, b), \text{on}(b, c), \text{green}(a), \sim\text{green}(c)\}$$

The question is whether there are two blocks x and y such that

$$X \vdash \text{on}(x, y) \wedge \text{green}(x) \wedge \sim\text{green}(y)$$

Clearly, the answer according to classical logic is 'yes': either b is green, then it is a green block on top of the nongreen block c, or b is not green, then a is a green block on top of the nongreen block b. This line of reasoning is based on the classical principle *tertium non datur* which implies that all predicates are exact, thus providing the indefinite information that $green(b) \vee \sim green(b)$ which in turn implies a positive answer to the above query. It is nonconstructive in the sense that it confirms an existential sentence without really being able to tell for which objects it holds (unless we are prepared to admit indefinite objects). The principle *tertium non datur* also does not leave room for the epistemologically motivated possibility of inexact predicates.

From the point of view of vivid logic, $green(b) \vee \sim green(b)$ does not hold tautologically. If we assume the predicate 'green' to be inexact[3] we have absolutely no information about the greenness of b, so we can neither conclude that it is green nor that it is not, hence we obtain the answer: no, it is not the case that there is a green block on top of a nongreen one since this cannot be proved for any pair of blocks.

If we assume the predicate 'green' to be exact we still have to distinguish between two cases. Either it is totally or partially represented. If it is totally represented we can apply the CWA rule to obtain the conclusion that b is not green, and hence the answer: yes, there is a green block on top of a nongreen one, namely a on top of b. If 'green' is exact but not totally represented we might add the disjunctive formula $green(b) \vee \sim green(b)$ as an indefinite information about the greenness of b to the database. This, together with an appropriate semantics of disjunctive databases allowing for indefinite answers, would allow to obtain the desired positive conclusion even in a nonclassical framework. I will only briefly sketch such a framework with the help of a further example. As already noticed above, legally defined predicates, such as *married* can be considered as exact (as opposed to empirical predicates, such as *green*). So it may make sense to require from the following knowledge base,

$$\text{KB} = \left\{ \begin{array}{l} \text{LookingAt(Mary, Peter)} \\ \text{LookingAt(Peter, Susan)} \\ \text{married(Mary)} \\ \sim\text{married(Susan)} \\ \{\text{married}(x) \vee \sim\text{married}(x) : x \in \{\text{Mary, Susan, Peter}\}\} \end{array} \right.$$

the conclusion that there is a married person looking at an unmarried one,

$$\text{KB} \vdash Q(x, y)$$

---

[3] In the philosophical literature there is indeed a strong tendency to qualify colour predicates as inexact, see e.g. [Kutschera 1984].

where

$$Q(x, y) = \text{LookingAt}(x, y) \wedge \text{married}(x) \wedge \sim \text{married}(y)$$

In order to accommodate indefinite answers the semantics of the existential quantifier and the related semantics of open queries has to be modified:

$$\text{KB} \vdash F(x) \quad \text{iff} \quad \text{KB} \vdash \bigvee_{i=1}^{n} F(t_i) \quad \text{for some } t_1, \ldots, t_n$$

where $\{t_1, \ldots, t_n\}$ is an answer, that is, each t_i is a tuple of ground terms. In the above example we get

$$\text{KB} \vdash Q(\text{Mary, Peter}) \vee Q(\text{Peter, Susan})$$

i.e. the indefinite answer

$$\{\langle \text{Mary, Peter} \rangle, \langle \text{Peter, Susan} \rangle\}$$

The dogma of bivalence and the resulting implicit indefiniteness of all information (which requires a nonconstructive interpretation of disjunction and existence) in classical logic expresses a notion of truth which is certainly not adequate for the purpose of vivid reasoning.

Partiality also occurs in relational databases. The RDB model handles unknown or not-yet-assigned attribute values correctly by assigning special *null* values.[4] In the case of a logical attribute *null* represents the truth-value **unknown**. If such an attribute corresponds to an exact predicate (which could be stipulated by legal definition in the case of an administrative database), the value *null* should count as 'no', i.e. the CWA rule could be applied in order to check falsity.

In comparison with [Levesque 1988] there are two main differences in the approach presented here:

1. Only certain predicates, and not all, are subject to the CWA. Therefore a KB is not, in general, complete (with respect to $\sim$).

2. Consistency is not required. A contradiction in the KB is not so bad (as it is in classical logic). Though an inconsistent KB does not have a proper partial model, it has a general one.

4.6 Epistemic States and Weak Negation

The basic VKRS made up by Belnap's epistemic states can easily be extended by adding weak negation (and an appropriate form of the CWA). In fact, there

---

[4] Cf. Codd [1970].

is no need to restrict the query language to formulas without weak negation. By definition (in 3.2.2) the derivability of a formula from an epistemic state Y is based on the model relation $\models \; \subseteq 2^{\text{Lit}} \times L(1, -, \sim, \wedge, \vee)$, therefore it is also defined for weakly negated query formulas:

Definition 4.7 $Y \vdash -F \overset{def}{\Longleftrightarrow} \quad$ *for all* $M \in Y : M \models -F$

Example 4.2 $\{\{p\}, \{q\}\} \vdash (p \vee q) \wedge -(p \wedge q)$. *However, neither p nor q, and neither $-p$ nor $-q$ are derivable.*

Observation 4.8 $Y \vdash F$ *iff* $\forall M \in Y \, \exists E \in \text{DNS}(F) : E^+ \subseteq M \; \& \; E^- \cap M = \emptyset$

While the inference relation can be simply extended in order to acommodate weak negation, this is not possible for the update operation the definition of which has to be changed in the case of disjunctive updates. According to the definition of $(U\vee)$ in 3.2.2, one would obtain that

$$\mathbf{Upd}(0, p \vee q) \vdash -p \vee -q$$

i.e. a disjunctive input would be interpreted as 'exclusive' whereas standard disjunction is rather 'inclusive'. Therefore the following new definition is needed:

$$(U\vee) \qquad \mathbf{Upd}(\text{KB}, F \vee G) \; := \; \mathbf{Upd}(\text{KB}, F) \cup \mathbf{Upd}(\text{KB}, G)$$
$$\cup \mathbf{Upd}(\text{KB}, F \wedge G)$$

For sets of extended literals as input, $E \subseteq \text{XLit}$, update is defined as

$$\mathbf{Upd}(Y, E) := \{X \cup E^+ - E^- : X \in Y\}$$

Using this definition, the update operation can also be formulated in terms of the disjunctive normal form of an input formula $F \in L(1, -, \sim, \wedge, \vee)$ in the following way:

$$\mathbf{Upd}(Y, F) = \bigcup_{Z \subseteq \text{DNS}(F)} \mathbf{Upd}(Y, \bigcup Z)$$

Notice that by using all subsets of the disjunctive normal set of an input formula F, **Upd** becomes exponential in the size of $\text{DNS}(F)$, making the amount of disjunctiveness of an input formula a computationally critical factor for updates.

Observation 4.9 $Y \vdash F \Rightarrow Y \subseteq \mathbf{Upd}(Y, F)$

Definition 4.8   *The KRS*

$$\langle 2^{2^{\text{Lit}}}, \vdash, L(-, \sim, \wedge, \vee), \text{Upd}, L(-, \sim, \wedge, \vee) \rangle$$

is denoted by V_B^- .

Claim 4.3 V_B^- *is a non-cumulative VKRS. It does not satisfy Lemma Compatibility.*

Proof: (Restricted Reflexivity) Let F be weakly consistent, and let $Y' = \text{Upd}(Y, F)$. It has to be shown that $Y' \vdash F$, i.e. if $M' \in Y'$ then $\mathcal{M}' \models F$. Since $M' = M \cup E^+ - E^-$ for some $M \in Y$ and some $E \in \text{DNS}(F)$ such that $E^+ \cap E^- = \emptyset$, it follows that $E^+ \subseteq M'$ and $E^- \cap M' = \emptyset$, i.e. $\mathcal{M}' \models F$.

(Lemma Redundancy) Let $Y \vdash F$ and $\text{Upd}(Y, F) \vdash G$. It has to be shown that $Y \vdash G$, or equivalently, if $M \in Y$ then $E^+ \subseteq M$ and $E^- \cap M = \emptyset$ for some $E \in \text{DNS}(G)$. Now, $\text{Upd}(Y, F) \vdash G$ implies that $E^+ \subseteq M'$ and $E^- \cap M' = \emptyset$ for some $E \in \text{DNS}(G)$ and all $M' \in \text{Upd}(Y, F)$, consequently also for all $M \in Y \subseteq \text{Upd}(Y, F)$.

(Constructivity) and (Non-Explosiveness) are established in the same way as in 3.2.2.

(Failure of Cumulativity, resp. Lemma Compatibility) A simple example demonstrates that lemmas are not always compatible:

$$\{\{p\}\} \vdash p \vee q$$
but $\text{Upd}(\{\{p\}\}, p \vee q) = \{\{p\}, \{p, q\}\} \not\vdash -q$
while $\{\{p\}\} \vdash -q$ □

Question 3   *Somewhat disturbing is the following example:*

$$\text{Upd}(\{\{p\}\}, p \vee -p) = \{\{p\}, \{\}\} \not\vdash p$$

where the consequence p is lost after the KB has been updated with the lemma $p \vee -p$. Strange things can happen .... Is this a feature or a bug ?

Besides Lemma Compatibility, also persistence and ampliativeness get lost by the addition of weak negation to Belnap's KRS.

Observation 4.10 $\vdash$ *is not persistent, in general. However, query formulas $F \in L(1, \sim, \wedge, \vee)$ are persistent.*

Observation 4.11   Upd *is no longer ampliative, that is, $Y \not\subseteq \text{Upd}(Y, F)$, in general. However, input formulas $F \in L(1, \sim, \wedge, \vee)$ are ampliative.*

Claim 4.4   *Epistemic states are unique representations with respect to query formulas $F \in L(-, \sim, \wedge, \vee)$, that is*

$$C(Y_1) = C(Y_2) \Rightarrow Y_1 = Y_2$$

Proof: For $i = 1, 2$ let

$$F_i = \bigvee_{M \in Y_i} (\bigwedge M \wedge \bigwedge \overline{\mathrm{Lit} - M})$$

be the characteristic formula of $Y_i = \mathbf{Upd}(\{\emptyset\}, F_i)$. Since F_i is weakly consistent, $Y_i \vdash F_i$, by Restricted Reflexivity. In order to prove the assertion it suffices now to show that $Y_1 \neq Y_2$ implies $Y_1 \nvdash F_2$ or $Y_2 \nvdash F_1$. Suppose that $Y_1 \vdash F_2$, i.e. $\forall M \in Y_1 \exists E \in \mathrm{DNS}(F_2) : E^+ \subseteq M$ & $E^- \cap M = \emptyset$, consequently by definition of F_2, $\forall M \in Y_1 \exists M' \in Y_2 : M' \subseteq M$ & $(\mathrm{Lit} - M') \cap M = \emptyset$, in other words, $Y_1 \subseteq Y_2$, since $(\mathrm{Lit} - M') \cap M = \emptyset$ iff $M \subseteq M'$. Similarly, if $Y_2 \vdash F_1$ then $Y_2 \subseteq Y_1$. $\square$

Chapter 5

Lindenbaum-Algebraic Semantics of Logic Programs

I show how to obtain the Lindenbaum algebra of a logic program without negation-as-failure. In the case of a positive program it is simply a distributive lattice with a greatest element. I also investigate programs with strong negation which allow to represent and process explicit negative information. Although these programs have double negation elimination and the DeMorgan rules one does not obtain a DeMorgan algebra as the Lindenbaum algebra of a program with strong negation as one could have expected.

5.1 Introduction

Logic programs are syntactically a fragment of first order logic, but semantically they must be viewed as a weakening of classical logic because they do not have classical negation behaving as a Boolean complement and giving rise to the nonconstructive behaviour of disjunction and conjunction observable in classical logic.

The logic of positive programs is subclassical. In fact, it can be viewed as the $\{\wedge, \vee\}$-fragment of positive logic which deals with $\{\wedge, \vee, \rightarrow\}$. Thus, it is not surprising that positive programs induce a distributive lattice as their underlying Lindenbaum-algebraic structure.

The approach taken here is to investigate the deductive behaviour of logic programs proof-theoretically – as is also done e.g. in [Miller 1989]. As the starting point a derivability relation between a program and a formula is defined. It captures the 'query evaluation procedure' actually carried out by the Prolog interpreter in order to show the relative validity of a formula. Then, following the general approach of Rasiowa [1974], a preordering on the set of all formulas based on the notion of derivability introduced in the first step is defined. Such an ordering leads to a set of equivalence classes which can be expected to be the Lindenbaum algebra looked for if the logical operators are monotone with respect to the preorder, and consequently, the induced equivalence relation is compatible with them.

While Rasiowa considers only systems with an appropriate implication by means of which she defines the preorder, I have to manage without such an implication. The construction presented here is based on the concept of *conditional derivability* which corresponds to first-degree implication.

The Lindenbaum algebra does not only provide an adequate model of a program by assigning truth to exactly those formulas which are derivable. It also gives a clear picture of the deductive structure induced by the program by eliminating all syntactic redundancy. This is expressed by means of the resulting ordering of all (congruence classes of) formulas.

While an algebraic semantics of the 'weak' negation operation '−' is still a topic of future research I investigate the possibility of an algebraic characterization of logic programs with strong negation in section 3. Although it is not obvious at all what algebraic structure would model those negation operations and their interaction with conjunction and disjunction it is clear that one will not end up with Boolean algebras.

Logic programs with a second negation, in addition to negation-as-failure, have only recently been proposed by several authors.[1] Yet, in these papers, the proposed negation is not classified and named properly.

Although only propositional logic programs will be considered in the sequel, the results are also applicable to logic programs in general by transforming them into their Herbrand expansion, or instantiation.

5.2 Positive Logic Programs

The language of positive logic programs consists of the operator symbols $\wedge$, $\vee$ and the constant symbol 1. A program *clause* is an expression of the form $a \leftarrow F$, where a is an atom and F a formula. For $F = 1$ the clause is called *fact*, otherwise *rule*[2]. A *logic program* (for short, *program*) Π is a finite set of clauses. Facts, $a \leftarrow 1$, are also abbreviated by a. At_Π denotes the set of all atoms occuring in a program Π.

As an example for a program clause take $a \leftarrow (b \wedge c) \vee d$ whose standard

---

[1] See [Gelfond & Lifschitz 1990], [Przymusinski 1990], [Kowalski & Sadri 1990].

[2] I do not consider program clauses as axioms, i.e. formulas, but rather as specific inference rules.

Prolog version would be a :- b,c;d.

5.2.1   The Atom Hierarchy of a Program

For a given program Π the single-step atomic consequence operation $C_\Pi^1 : 2^{\mathrm{At}_\Pi} \to 2^{\mathrm{At}_\Pi}$ is defined as:[3]

$$C_\Pi^1(X) \quad := \quad \{\, a \in \mathrm{At}_\Pi : \text{there is } a \leftarrow F \in \Pi \text{ such that } X \vdash F \,\}$$

where $X \vdash F$ is an inference on the basis of facts as described in 4.1. The atom hierarchy of a program is obtained as follows:

$$\Pi^0 := \emptyset, \quad \Pi^{i+1} := C_\Pi^1(\Pi^i), \quad \Pi^\omega := \bigcup_{i<\omega} \Pi^i$$

Obviously, Π^ω contains all atoms derivable from Π by the successive application of C_Π^1.

Observation 5.1 C_Π^1 *is isotone, i.e.*

$$X \subseteq Y \;\Rightarrow\; C_\Pi^1(X) \subseteq C_\Pi^1(Y)$$

It follows that $\Pi^i \subseteq \Pi^{i+1}$ for all $i < \omega$ and $\Pi^\omega = \Pi^n$ for a certain $n < \omega$. This leads to the definition of derivability from a program:

$$
\begin{array}{lll}
(1) & \Pi \vdash 1 & \\
(a) \quad \text{if} & a \in \Pi^\omega & \\
(\wedge) \quad \text{if} & \Pi \vdash F \text{ and } \Pi \vdash G & \\
(\vee) \quad \text{if} & \Pi \vdash F \text{ or } \Pi \vdash G &
\end{array}
$$

Notice that derivability from Π corresponds to derivability from Π^ω:

$$\Pi \vdash F \quad \text{iff} \quad \Pi^\omega \vdash F$$

There are no nontrivial tautologies, i.e. no formulas F such that for every program Π, $\Pi \vdash F$ (trivial cases contain 1 like, for example, $1 \vee F$).

Essentially, this notion of derivability captures the 'query evaluation procedure' of positive Prolog. Of course, Prolog does not do the bottom-up computation of Π^ω but computes the derivability of atoms in a top-down fashion which might not terminate in case of dependency loops, thus, rendering the right hand side of

$$(a') \quad \Pi \vdash a \quad \text{iff} \quad a \leftarrow F \in \Pi \,\&\, \Pi \vdash F$$

---

[3]C_Π^1 is just the propositional logic version of the now classical notion of a monotone program operator T_Π leading to the fixed point semantics of van Emden and Kowalski.

undecided. For certain programs, however, which are called *wellfounded* according to the definition 6.1 in 6.5, (a) can be replaced by (a').

Recall the disjunctive normal form of a formula G,

$$\mathrm{DNF}(G) = \bigvee_{K \in \mathrm{DNS}(G)} \bigwedge K$$

which is equivalent to G according to the following

Observation 5.2 $\Pi \vdash G$ *iff* $\Pi \vdash \mathrm{DNF}(G)$

Proof by induction on G: it is easy to show that the assertion holds for $G = a$ and $G = H \vee H'$. In the remaining case, $G = H \wedge H'$, we have $\Pi \vdash \mathrm{DNF}(H \wedge H')$ iff $\Pi \vdash \bigvee \bigwedge K$ where K ranges over all sets $H_i \cup H_j'$ with $H_i \in \mathrm{DNS}(H)$ and $H_j' \in \mathrm{DNS}(H')$. This holds iff there is a pair $H_i \in \mathrm{DNS}(H)$ and $H_j' \in \mathrm{DNS}(H')$ such that $\Pi \vdash \bigwedge(H_i \cup H_j')$ which is equivalent to $\Pi \vdash \bigwedge H_i$ and $\Pi \vdash \bigwedge H_j'$, consequently also to $\Pi \vdash \mathrm{DNS}(H)$ and $\Pi \vdash \mathrm{DNS}(H')$. By the induction hypothesis this is equivalent with $\Pi \vdash H$ and $\Pi \vdash H'$, and consequently with $\Pi \vdash H \wedge H'$. $\square$

5.2.2   Valuations in Distributive Lattices as Models

Definition 5.1 *A valuation[4] v in a distributive lattice A with greatest element 1, $\langle A, \wedge, \vee, 1 \rangle$, is called a model of a program Π, if for every clause $a \leftarrow F \in \Pi$,*

$$v(F) = 1 \;\Rightarrow\; v(a) = 1$$

The following theorem establishes that derivability from a program is sound with respect to such models.

Claim 5.1   *For all models v of Π the following is valid:* $\Pi \vdash F \;\Rightarrow\; v(F) = 1$.

Proof by simultaneous induction on the exponent of Π^n and the complexity of F:

In the atomic case, $F = a$, it has to be shown that $a \in \Pi^\omega$ implies $v(a) = 1$. If $a \in \Pi^1$ then there is $a \leftarrow 1 \in \Pi$, and since v is a model this implies $v(a) = 1$. Now let $a \in \Pi^n$. Then, there is a clause $a \leftarrow F \in \Pi$ such that $\Pi^{n-1} \vdash F$. I show that $v(F) = 1$, and consequently $v(a) = 1$, by induction on F. This is clear for $F = 1$. If $F = b$ the n-induction hypothesis implies $v(b) = 1$ since $b \in \Pi^{n-1}$. For $F = G \wedge H$ we get $\Pi^{n-1} \vdash G$ and $\Pi^{n-1} \vdash H$, consequently by the F-induction hypothesis, $v(G) = 1$ and $v(H) = 1$ implying $v(G \wedge H) = 1$. Similarly for $F = G \vee H$.

It remains to show by straightforward induction that the assertion holds in the complex cases, $F = G \wedge H$ and $F = G \vee H$, too. $\square$

----

[4] A valuation in an algebra A is a homomorphism from the set of formulas into A, see also [Rasiowa 1974].

5.2.3 Construction of an Adequate Model

Now a ternary derivability relation between a program, a premise formula and a conclusion formula is defined:

$$\Pi, F \vdash G \quad \text{iff} \quad \text{for every } K \in \text{DNS}(F) : \Pi \cup K \vdash G$$

This relation captures the notion of positive implication between (implication-free) formulas in the sense of the metalogical implication $\Pi \vdash F \Rightarrow \Pi \vdash G$. Indeed, as a corollary of observation 5.7, the *cut* rule is obtained:

$$\Pi \vdash F \ \& \ \Pi, F \vdash G \Rightarrow \Pi \vdash G$$

As an abbreviation, I also write Π_K for $\Pi \cup K$. For the sake of simplicity, I will sometimes identify the set of atoms K with the corresponding conjunction $\bigwedge K$.

The proof of the following two observations is omitted:

Observation 5.3   *1.* $\Pi, F \vdash G$ *iff* $\Pi, \text{DNF}(F) \vdash G$

2. $\Pi, F \vdash G$ *iff* $\Pi, F \vdash \text{DNF}(G)$

Notice that, as a consequence of this observation, conjunction and disjunction are commutative in the premise as well as in the conclusion.

Observation 5.4   *1.* $\Pi, F \vdash G \wedge H$ *iff* $\Pi, F \vdash G \ \& \ \Pi, F \vdash H$

2. $\Pi, F \vdash H \Rightarrow \Pi, F \wedge G \vdash H$

3. $\Pi, F \vdash H \Rightarrow \Pi, F \vdash H \vee G$

4. $\Pi, F \vee G \vdash H$ *iff* $\Pi, F \vdash H \ \& \ \Pi, G \vdash H$

Observation 5.5 $\Pi, F \vdash F$.

Proof by induction on F: In the atomic case the assertion is obvious. Let $F = G \vee H$. From the induction hypothesis it follows with observation 5.4.3, that $\Pi, G \vdash G \vee H$ and $\Pi, H \vdash G \vee H$, hence the assertion by observation 5.4.4. From the induction hypothesis for $\wedge$ it follows with observation 5.4.2, that $\Pi, G \wedge H \vdash G$ and $\Pi, G \wedge H \vdash H$, hence the assertion by observation 5.4.1. $\square$

Observation 5.6   *Let F and G be conjunctions of atoms or sets of facts, respectively. Then the following holds: $G \subseteq \Pi_F^\omega \Rightarrow \Pi_G^\omega \subseteq \Pi_F^\omega$ (where $\Pi_F^\omega = (\Pi \cup F)^\omega$)*

Proof by induction on the exponent of Π_G^n: In the base case, $n = 0$, we trivially have $\Pi_G^0 = \emptyset \subseteq \Pi_F^\omega$. Now, assume that $\Pi_G^k \subseteq \Pi_F^\omega$ for all $k < n$ given the proviso. Then for $a \in \Pi_G^n$ we obtain two cases: either $a \in G$, and hence, $a \in \Pi_F^\omega$ by the proviso, or $a \leftarrow H \in \Pi$ and $\Pi_G^{n-1} \vdash H$ which implies by the induction hypothesis and the monotonicity of $\vdash$ that $\Pi_F^\omega \vdash H$, and hence, $a \in \Pi_F^\omega$. $\square$

Observation 5.7 $\Pi, F \vdash G$ & $\Pi, G \vdash H \Rightarrow \Pi, F \vdash H$

Proof: By observation 5.3, $\Pi, F \vdash G$ implies $\Pi_{F_i} \vdash G_{k(i)}$ for all $i = 1 \dots m$
(where $\mathrm{DNF}(G) = \bigvee G_k$ and k(i) is the specific value of k for which $\Pi_{F_i} \vdash G_k$).
In other words, $G_{k(i)} \subseteq \Pi_{F_i}^\omega$ for all i. This implies $\Pi_{G_{k(i)}}^\omega \subseteq \Pi_{F_i}^\omega$ for all i by
observation 5.6. Since we also have $\Pi_{G_k} \vdash H$ for all k by the second proviso, the
assertion follows by the monotonicity of $\vdash$. $\square$

Claim 5.2   *A program determines a preorder on the set of all formulas.*

Proof: By observation 5.5 and 5.7, the relation

$$F \leq_\Pi G \stackrel{def}{\Longleftrightarrow} \Pi, F \vdash G$$

is reflexive and transitive, i.e. a preorder. $\square$

As is well-known, a preorder $\leq_\Pi$ leads to an equivalence relation:

Definition 5.2 $F \simeq_\Pi G \stackrel{def}{\Longleftrightarrow} F \leq_\Pi G$ & $G \leq_\Pi F$

Claim 5.3 $\simeq_\Pi$ *is a congruence relation.*

Proof: It suffices to show that for $\star = \wedge, \vee$

$$F_1 \leq_\Pi G_1 \ \& \ F_2 \leq_\Pi G_2 \Rightarrow F_1 \star F_2 \leq_\Pi G_1 \star G_2$$

which follows straightforwardly from observation 5.4. $\square$

This means we can form the corresponding quotient algebra of formulas modulo Π-equivalence, also called *Lindenbaum algebra*. Let L_Π be the set of all
formulas in $L(1, \wedge, \vee)$ generated by At_Π and let $\mathcal{F}_\Pi = L_\Pi / \simeq_\Pi$.

Claim 5.4   *The canonical valuation $v_\Pi : L_\Pi \to \mathcal{F}_\Pi$ with $v_\Pi(a) = a / \simeq_\Pi$, and
consequently $v_\Pi(F) = F / \simeq_\Pi$, is an adequate model of Π, i.e.*

$$\Pi \vdash F \quad \textit{iff} \quad v_\Pi(F) = 1 / \simeq_\Pi$$

Proof: First, it is shown that v_Π is a model of Π. It is easy to verify that the
Lindenbaum algebra $\mathcal{F}_\Pi$ is a distributive lattice. For each clause $a \leftarrow F \in \Pi$,
$F \leq_\Pi a$ by definition, and as a consequence, $\overline{F} \leq \overline{a}$ in the Lindenbaum algebra
where I abbreviate $\overline{F} := F / \simeq_\Pi$. Then, if $v_\Pi(F) = \overline{1}$ the value of a must be
greater or equal than that of F, consequently it must be $\overline{1}$, as well. Finally, v_Π
is adequate since $v_\Pi(F) = \overline{1}$ implies that $\Pi, 1 \vdash F$, hence $\Pi \vdash F$. $\square$

The Lindenbaum algebra of a positive logic program Π can as well be described as the distributive lattice freely generated by At_Π modulo all inequalities
$F \leq a$ where $a \leftarrow F \in \Pi$.

5.2.4   Example

For the following program

$$\Pi \;=\; \begin{cases} a \leftarrow b \\ a \leftarrow 1 \\ b \leftarrow a \wedge c \\ c \leftarrow b \\ d \leftarrow (b \wedge a) \vee c \end{cases}$$

we obtain a three element algebra base $\text{At}_\Pi/\simeq_\Pi$ ordered as $\{b,c\} \le \{d\} \le \{1,a\}$ which generates $\mathcal{F}_\Pi$:

$$\mathcal{F}_\Pi \;=\; \begin{array}{c} 1 \\ | \\ \overline{d} \\ | \\ \overline{b} \end{array}$$

5.2.5   Possible Applications

As possible applications of the above ideas the following spring to mind:

1. Two logic programs Π and Π' can be considered to be *logically equivalent* if both have the same Lindenbaum algebra: $\mathcal{F}_\Pi = \mathcal{F}_{\Pi'}$.

2. One can introduce an implication operator in the language of logic programs by means of the introduction rule $\Pi \vdash F \rightarrow G \overset{def}{\Longleftrightarrow} \Pi, F \vdash G$.

3. The Lindenbaum algebra of a logic program could be obtained by compilation and then be used to check valid deductions by means of checking ordering relations. For this purpose, *deductive filters* as described in [Rasiowa 1974] might be useful.

5.3   Logic Programs with Strong Negation

Now complex formulas are built from atomic ones with the help of the binary operators $\wedge$ and $\vee$, the unary operator $\sim$ and the constant 1. The constant expression ~ 1 is abbreviated by 0. A *literal* is either positive, i.e. an atom like a, or negative, i.e. a negated atom like $\sim a$. A program *clause* is an expression of the form $l \leftarrow F$, where the conclusion l is a literal and the premise F is a formula. The *literal base* Lit_Π of a program Π is defined as $\text{Lit}_\Pi = \text{At}_\Pi \cup \{\sim a : a \in \text{At}_\Pi\}$.

Logic programs in this sense correspond in a certain way to Prolog programs where the negation operator **not** may occur only in the premise of a clause. So,

$$a \leftarrow b \wedge (c \vee \sim d)$$

corresponds syntactically to **a :- b, (c; not d)** , whereas $\sim a \leftarrow \sim b \vee c$ has no equivalent in Prolog syntax. In terms of semantics there is even a greater disagreement: negation-as-failure is not a 'strong' but rather a 'weak' form of negation. So, by introducing strong negation into logic programming I do not claim to provide the logical semantics of **not**. However, as argued in [Pearce & Wagner 1989], the incorporation of strong negation in logic programs is not only interesting from a theoretical point of view but also promising in terms of practical applications. Note that strong negation does not oust negation-as-failure but rather supplement it.

5.3.1   The Literal Hierarchy of a Program

For a given program Π and $X \subseteq \text{Lit}_\Pi$ let the single-step literal consequence operation $C_\Pi^1(X)$ be defined as

$$C_\Pi^1(X) \quad := \quad \{ l \in \text{Lit}_\Pi : \text{ there is } l \leftarrow F \in \Pi, \text{ such that } X \vdash F \}$$

C_Π^1 is an isotone mapping from 2^{Lit_Π} to 2^{Lit_Π}. By means of it 'powers' of a program describing a hierarchy of derivable literals can be defined:

$$\Pi^0 := \emptyset, \quad \Pi^{i+1} := C_\Pi^1(\Pi^i), \quad \Pi^\omega := \bigcup_{i < \omega} \Pi^i$$

Obviously, Π^ω contains all literals derivable from Π by the successive application of C_Π^1. Since C_Π^1 is isotone, $\Pi^i \subseteq \Pi^{i+1}$ for all $i < \omega$ and $\Pi^\omega = \Pi^n$ for a certain $n < \omega$.

This leads to the definiton of derivability from a program:

$$
\begin{array}{lll}
(1) & & \Pi \vdash 1 \\
(l) & \text{if} & l \in \Pi^\omega \\
(\wedge) & \text{if} & \Pi \vdash F \text{ and } \Pi \vdash G \\
(\sim\wedge) & \text{if} & \Pi \vdash \sim F \text{ or } \Pi \vdash \sim G \\
(\sim\sim) & \text{if} & \Pi \vdash F
\end{array}
$$

The rules for disjunction ($\vee$) and ($\sim\vee$) can be derived.

Note that derivability from Π corresponds to derivability from the set of literals induced by Π, Π^ω: $\Pi \vdash F$ iff $\Pi^\omega \vdash F$.

In general, it is not valid that $\Pi \vdash F \vee \sim F$, hence also not $\Pi \vdash \sim (F \wedge \sim F)$. There are no nontrivial tautologies, i.e. no formulas F such that for every program Π, $\Pi \vdash F$ (trivial cases being formulas containing 1 like $1 \vee F$).

For the proof of the following observation see the corresponding observation 5.2 in the previous section.

Observation 5.8 $\Pi \vdash G$ *iff* $\Pi \vdash \mathrm{DNF}(G)$

5.3.2   Valuations in DeMorgan Algebras

A *DeMorgan algebra* is a distributive lattice with a DeMorgan complement, i.e. a unary operation satisfying the law of double negation and the DeMorgan rules. Formally, $\langle A, \wedge, \vee, \sim, 1 \rangle$ is a DeMorgan algebra with greatest element 1 if

(1) $\langle A, \wedge, \vee, 1 \rangle$ is a distributive lattice with greatest element 1

(2) $\sim \sim a = a$, and $\sim (a \vee b) = \sim a \wedge \sim b$

It follows that

(3) $\sim (a \wedge b) = \sim a \vee \sim b$

(4) $a \leq b \Rightarrow \sim b \leq \sim a$

(5) $0 := \sim 1$ is the least element

A valuation v in a DeMorgan algebra with greatest element 1 is called a model of a program Π, if for every clause $l \leftarrow F \in \Pi$, $v(F) = 1 \Rightarrow v(l) = 1$.

Derivability from a program is sound with respect to DeMorgan algebraic models:

Claim 5.5   *For all models v of Π, $\Pi \vdash F \Rightarrow v(F) = 1$.*

Proof by simultaneous induction on the exponent of Π^n and the complexity of F: In the literal case, $F = l$, it has to be shown that $l \in \Pi^\omega$ implies $v(l) = 1$. If $l \in \Pi^1$ then there is $l \leftarrow 1 \in \Pi$, and since v is a model this implies $v(l) = 1$. Now let $l \in \Pi^n$. Then, there is a clause $l \leftarrow F \in \Pi$ such that $\Pi^{n-1} \vdash F$. I show that $v(F) = 1$, and consequently $v(l) = 1$, by induction on F. This is clear for $F = 1$. In the case of a literal, $F = k$, the n-induction hypothesis implies $v(k) = 1$ since $k \in \Pi^{n-1}$. For $F = G \wedge H$ we get $\Pi^{n-1} \vdash G$ and $\Pi^{n-1} \vdash H$, consequently by the F-induction hypothesis, $v(G) = 1$ and $v(H) = 1$ implying $v(G \wedge H) = 1$. For $F = \sim (G \wedge H)$ we get $\Pi^{n-1} \vdash \sim G$ or $\Pi^{n-1} \vdash \sim H$, consequently by the F-induction hypothesis $v(\sim G) = 1$ or $v(\sim H) = 1$, implying that $v(\sim (G \wedge H)) = v(\sim G \vee \sim H) = 1$. Finally, $v(\sim \sim G) = 1$ if $v(G) = 1$. It remains to show by straightforward induction that the assertion holds in the general complex cases, too. □

5.3.3   Construction of an Adequate Model

For logic programs with strong negation the relation $\Pi, F \vdash G$ is defined in the same way as for positive programs. Likewise, observation 5.3 holds, stating that both the premise formula and the conclusion formula of the derivability relation $\vdash$ can be normalized. This implies that double negation can be eliminated on both sides of $\vdash$.

Observation 5.9   *1.* $\Pi, F \vdash \sim G \; \Rightarrow \; \Pi, F \vdash \sim(G \wedge H)$

   2. $\Pi, \sim G \vdash F \; \& \; \Pi, \sim H \vdash F \; \Rightarrow \; \Pi, \sim(G \wedge H) \vdash F$

In order to obtain an equivalence relation between formulas compatible with $\sim$ a preorder on the set of all formulas is defined in a similar way as in [Rasiowa 1974]:

$$F \leq_\Pi G \; \stackrel{def}{\Longleftrightarrow} \; \Pi, F \vdash G \text{ and } \Pi, \sim G \vdash \sim F$$

Notice that this definition implies that in order to have $F \leq_\Pi 1$ one has to stipulate that $\Pi, 0 \vdash F$ which is nothing else as the well-known principle *ex falso sequitur quodlibet*. However, with this ordering the canonical valuation in $\mathcal{F}_\Pi$, in general, is not a model of Π. Before discussing this problem I show that $\leq_\Pi$ is indeed a preorder, i.e. a reflexive and transitive relation.

Observation 5.10 $F \leq_\Pi F$, *i.e.* $\leq_\Pi$ *is reflexive.*

Proof by induction on F: I just treat the negative subcase of $F = G \wedge H$ and show that $\Pi, \sim(G \wedge H) \vdash \sim(G \wedge H)$. From the induction hypothesis it follows that $\Pi, \sim G \vdash \sim G$ and $\Pi, \sim H \vdash \sim H$. Thus, by observation 5.9.1, $\Pi, \sim G \vdash \sim(G \wedge H)$ as well as $\Pi, \sim H \vdash \sim(G \wedge H)$, consequently the assertion is obtained by observation 5.9.2. The remaining cases $F = \sim(G \wedge H)$ and $F = \sim\sim G$ follow from this and/or double negation elimination. $\square$

Observation 5.11   *Let F and G be conjunctions of literals or sets of facts, respectively. Then the following holds:* $G \subseteq \Pi_F^\omega \; \Rightarrow \; \Pi_G^\omega \subseteq \Pi_F^\omega$ *(where* $\Pi_F^\omega = (\Pi \cup F)^\omega$ *)*

Proof: Besides the fact that one deals with literals now instead of atoms the proof is the same as in the corresponding observation 5.6. $\square$

Observation 5.12 $\leq_\Pi$ *is transitive.*

Proof: It is easy to see that it suffices to show that $\Pi, F \vdash G \; \& \; \Pi, G \vdash H \; \Rightarrow \; \Pi, F \vdash H$ the proof of which is the same as that of the corresponding observation 5.7. $\square$

   It is clear from observations 5.9–5.12 that already the relation $\Pi, F \vdash G$ is a preorder, and hence leads to an equivalence relation between formulas. The problem is that this equivalence relation is not compatible with $\sim$ but only with $\wedge$ and $\vee$. Consequently, the corresponding factorization of L_Π does not yield a DeMorgan algebra.

Claim 5.6 *The relation $\simeq_\Pi$ defined by*

$$F \simeq_\Pi G \overset{def}{\Longleftrightarrow} F \leq_\Pi G \ \& \ G \leq_\Pi F$$

is a congruence relation.

Proof: I show that $\wedge$ and $\vee$ are isotone and $\sim$ is antitone with respect to $\leq_\Pi$ which implies the compatibility of $\simeq_\Pi$ with those operators.

($\sim$) The fact that $F \leq_\Pi G$ implies $\sim G \leq_\Pi \sim F$ (and vice versa) is a straightforward consequence of the definition of $\leq_\Pi$ and double negation elimination.

($\wedge$) Suppose we have $F_i \leq_\Pi G_i$ (for $i = 1, 2$), i.e. the following derivability relations hold: $\Pi, F_i \vdash G_i$ & $\Pi, \sim G_i \vdash \sim F_i$. From the former one it follows by observation 5.4 that $\Pi, F_1 \wedge F_2 \vdash G_1 \wedge G_2$. From the latter one it follows by observation 5.9 that $\Pi, \sim(G_1 \wedge G_2) \vdash \sim(F_1 \wedge F_2)$. Hence, $F_1 \wedge F_2 \leq_\Pi G_1 \wedge G_2$.

($\vee$) Disjunction, as a defined operation, is isotone, too:

$$
\begin{aligned}
F_1 \leq_\Pi G_1 \ \& \ F_2 \leq_\Pi G_2 \quad &\text{iff} \quad \sim G_1 \leq_\Pi \sim F_1 \ \& \ \sim G_2 \leq_\Pi \sim F_2 \\
&\text{iff} \quad \sim G_1 \wedge \sim G_2 \leq_\Pi \sim F_1 \wedge \sim F_2 \\
&\text{iff} \quad \sim(\sim F_1 \wedge \sim F_2) \leq_\Pi \sim(\sim G_1 \wedge \sim G_2) \\
&\text{iff} \quad F_1 \vee F_2 \leq_\Pi G_1 \vee G_2 \quad \square
\end{aligned}
$$

Let L_Π be the set of all formulas in $L(1, \sim, \wedge, \vee)$ generated by At_Π. A simple example shows that the canonical valuation in $\mathcal{F}_\Pi := L_\Pi/\simeq_\Pi$, in general, is not a model of Π. For $\Pi = \{a \leftarrow 1\}$ it is not the case that $1 \leq_\Pi a$, hence $v_\Pi(a) = a/\simeq_\Pi \neq 1$, although $\Pi \vdash a$.

Definition 5.3 Π *is called* contrapositionally complete *if for every clause* $l \leftarrow F \in \Pi$, $\Pi, \sim l \vdash \sim F$.

For example, $\Pi = \{a \leftarrow 1, \ 0 \leftarrow \sim a\}$ is contrapositionally complete.

Claim 5.7 *For a contrapositionally complete program Π the canonical valuation $v_\Pi : L_\Pi \to \mathcal{F}_\Pi$ with $v_\Pi(a) = a/\simeq_\Pi$, and consequently $v_\Pi(F) = F/\simeq_\Pi$, is an adequate model of Π, i.e.*

$$\Pi \vdash F \quad \text{iff} \quad v_\Pi(F) = 1/\simeq_\Pi$$

Proof: First, I show that v_Π is a model of Π. It is easy to verify that $\mathcal{F}_\Pi$ is a DeMorgan algebra. For each clause $l \leftarrow F \in \Pi$, $F \leq_\Pi l$ by the definition of $\leq_\Pi$ and the proviso that Π be contrapositionally complete. Consequently, $\overline{F} \leq \overline{l}$ in $\mathcal{F}_\Pi$ where $\overline{F} := F/{\simeq_\Pi}$. Then, if $v_\Pi(F) = \overline{1}$ the value of l must be greater or equal than that of F, consequently it must be $\overline{1}$, as well. Finally, v_Π is adequate since $v_\Pi(F) = \overline{1}$ implies that $\Pi, 1 \vdash F$, hence $\Pi \vdash F$. $\square$

The Lindenbaum algebra of a contrapositionally complete program Π can as well be described as the DeMorgan algebra freely generated by At_Π modulo all inequalities $F \leq l$ where $l \leftarrow F \in \Pi$.

Note that the contrapositional completion of a program, in general, is not a program. For example, the completion of $\Pi = \{a \leftarrow {\sim}b \wedge {\sim}c\}$ is

$$\{a \leftarrow {\sim}b \wedge {\sim}c,\ b \vee c \leftarrow {\sim}a\}$$

where the latter rule is not a program clause.

5.3.4 Example

For the following contrapositionally complete program

$$\Pi \quad = \quad \begin{cases} a \leftarrow b \vee {\sim}b \\ b \leftarrow {\sim}a \\ {\sim}b \leftarrow {\sim}a \end{cases}$$

one obtains

$$\mathrm{At}_\Pi/{\simeq_\Pi} \quad = \qquad \mathcal{F}_\Pi \quad =$$

5.4 Conclusion

I have shown that the notion of derivability from a program can be analyzed as a preorder which leads to a certain algebraic structure constructed by equivalence classes of formulas. This algebra can be considered as imposing constraints on the possible truth value assignments of models of the program.

In the case of positive programs a distributive lattice was obtained as the underlying algebraic structure. Although derivability from logic programs with strong negation was shown to be sound with respect to DeMorgan algebras those algebras did not prove to be the general model structure looked for. This is because there is no contraposition law for conditional derivability with respect to strong negation. Only for contrapositionally complete programs the construction of the Lindenbaum algebra proved to be adequate. In the general case it remains an open question what is the appropriate algebraic structure constituting the Lindenbaum algebra.

Chapter 6

Logic Programming with Strong Negation and Inexact Predicates

It is shown how a negation operation which allows for the possibility to represent explicit negative information can be added to the language of positive logic programs without altering the computational structure essentially. This negation is called *strong* since it expresses a notion of directly established falsity as opposed to the rather weak notion of indirectly established falsity expressed by negation-as-failure. By means of strong negation the useful distinction between exact and inexact predicates can be expressed in logic programs providing for the capability of reasoning with predicates from empirical domains.

6.1 Introduction

According to the standard view, a logic program is a set of definite Horn clauses. Thus, logic programs are regarded as syntactically restricted first order theories within the framework of classical logic. Correspondingly, the proof theory of logic programs is considered as the specialized version of classical resolution, known as SLD-resolution. This view, however, neglects the fact that a program clause, $a_0 \leftarrow a_1, a_2, \ldots, a_n$, is an expression of a fragment of positive logic[1] rather than an implicational formula of classical logic. The logical behaviour of such clauses is in no way related to any negation or complement operation. So, (positive) logic programs are 'subclassical'.[2] The classical interpretation seems

---

[1] A subsystem of intuitionistic logic.
[2] The algebraic structure underlying a positive logic program is a distributive lattice (and not a Boolean algebra) as is shown in 5.2.

to be a semantical overkill.

It should be clear that in order to explain the deduction mechanism of Prolog one does not have to refer to the indirect method of SLD-resolution which checks for the refutability of the contrary. It is certainly more natural to view Prolog's proof procedure as a kind of natural deduction, as, e.g., in [Hallnäs & Schroeder-Heister 1987] and [Miller 1989]. This also is more in line with the intuitions of a Prolog programmer. Since Prolog is the paradigm, logic programming semantics should take it as a point of departure. Therefore, I will only incidentally discuss issues of pure logic. Rather I will attempt to develop a logic programming semantics in the spirit of logic programming itself (which obviously is a constructive activity) and not by imposing some theoretical system borrowed from the literature on pure logic.[3]

With this in mind, I present here a system extending Prolog by adding strong negation which, in essential respects, is already known from constructive logics[4], and, in a limited form, from 3-valued, and from partial logic, respectively. This system is a conservative extension: it respects the top-down query evaluation procedure and the Herbrand fixed point semantics of Prolog. It can additionally be equipped with a negation-as-failure operator in the same way (creating the same problems) as positive standard Prolog. Below, I present a model and proof theory for the proposed logic programming system and discuss possible applications as well as its relation to other systems.

The work presented here is not inspired by relevance logic but rather by constructive logic which is a historic predecessor of relevance logic. Relevant implication is contrapositive, a property not wanted in logic programs. Also, here the relevance property in the notion of proof and the related semantics of implication is not wanted. The logic programming system proposed can be viewed as the *first-degree* fragment of Nelson's logic **N** (see 3.3). Therefore the usual possible worlds semantics for full constructive implication is not needed. A much simpler partial semantics will do it.

Akama [1987] has proposed to use constructive logic for the interpretation of logic programming. He shows how the resolution calculus for definite Horn clauses can be interpreted within constructive logic. In fact, he makes the point that in the definite Horn clause setting the differences between strong negation, intuitionistic negation and classical negation do not matter. This observation is not surprising since the language of definite Horn clauses can be viewed as a fragment of positive logic which forms a common subpart of constructive, intuitionistic and classical logic.

Most research on negation in logic programming has concentrated on the handling of *implicit negative information*, as embodied in negation-as-failure and

---

[3] However, I am not concerned with problems of unification.

[4] By *constructive logics* I mean systems essentially based on intuitionistic implication and strong negation (alias constructible falsity). Such systems have originally been proposed by Fitch [1948 + 1952], Nelson [1949] and Markov [1950].

related notions like Reiter's Closed-World Assumption and Clark's Completed Database. Some authors, notably Gabbay and Sergot [1986] and Miller [1989], have suggested to use *negation-as-inconsistency* in order to express negative information in the form of integrity constraints.

Negation-as-inconsistency, however, is less fundamental, since it can be defined by means of strong negation, as Nelson [1949] has shown.[5] In contrast to negation-as-inconsistency, strong negation is a local notion. Consequently, it is computationally affordable. Strong negation expresses constructible falsity which is established directly, whereas falsity in classical, intuitionistic and minimal logic is not a constructive notion and is typically established by an indirect proof. This also indicates that strong negation is the natural choice of negation for AI systems requiring the capability of processing explicitly represented negative information.

The AI community does not seem to be aware of the concept of strong negation, and one can often find confusion in the treatment of negation in the literature. For example, the negation operation **not** in Prolog is sometimes semantically treated as strong negation though it is rather a weak type of negation. Or one can find ad-hoc treatments of problems where strong negation would be needed. For instance, in [Levesque 1986] it is suggested to transform a negative literal like $\neg married(Jack, Jan)$ to an atom like $not\_married(Jack, Jan)$ in order to avoid the computational difficulties of classical negation. While this transformation is justified to a certain extent by a well-known theorem on the eliminability of strong negation[6] it destroys the contradiction relationship between a literal and its contrary in the object language. As a consequence of this, all reasoning capabilities based on inconsistency handling get lost.

The alternative suggested here is to express explicit negative information by means of strong negation. Just add the negative fact $\sim married(Jack, Jan)$ to the database. If also $married(Jack, Jan)$ can be inferred, the database is inconsistent, and one has to decide how to deal with it: for instance, accept both of the contradictory literals but no other purely 'inconsistency-based' conclusions, as in most paraconsistent logics, or conclude everything, as in logics based on the classical principle *ex contradictione sequitur quodlibet*. I will not pursue these questions in the present paper[7] but only illustrate them in 6.9.

Additionally, many empirical predicates which are to be represented e.g. in Prolog-based expert systems for diagnostic reasoning have to be considered as *inexact predicates* (see 4.3). It turns out that the semantical account for logic programs with strong negation given here is capable of handling inexact as well as exact predicates. This seems to emphasise the practical significance of the proposed system.

---

[5] See also 6.9.

[6] See 6.8.1. Gelfond and Lifschitz [1990] and Przymusinski [1990] make use of this, although they apparently are not aware of it.

[7] For a discussion of some new ideas related to inconsistency handling see [Wagner 1991c].

6.2 Logic Programs with Strong Negation

The language of logic programs with strong negation consists of the logical operator symbols $\wedge$, $\vee$, $\sim$ and 1 standing for conjunction, disjunction, strong negation and the verum, respectively, predicate symbols, constant symbols, function symbols and variables. Notice that there are no explicit quantifiers.

A program Π consists of *clauses* of the form $l \leftarrow F$ which are considered as specific inference rules and not as implicational formulas. A rule with premise 1 is also called a *fact*, and abbreviated by l. Examples of clauses are

$$\sim flies(x)emu(x) \vee penguin(x)$$
$$switch\_on\_light dark \wedge \sim illuminated$$

In 6.8.1, I will present a method by means of which a clause can be transformed into a set of expressions $a_0 \leftarrow a_1 \wedge \ldots \wedge a_n$ which are usually called 'definite Horn clauses'.[8] A logic program with strong negation is an extended positive program in the terminolgy of [Gelfond & Lifschitz 1990] as opposed to a normal program which allows for negation-as-failure in the premise of clauses.

A program Π containing non-ground clauses is considered as a dynamic representation of the corresponding set of ground clauses formed by means of the current domain of individuals U and denoted by $[\Pi]_U$. Formally,

$$[\Pi]_U := \{l\sigma \leftarrow F\sigma : l \leftarrow F \in \Pi \text{ and } \sigma : \mathrm{Var}(l, F) \to U\}$$

where σ ranges over all mappings from the set of variables of l and F into the Herbrand universe U. σ is called a *ground substitution* for $l \leftarrow F$ and $[\Pi]_U$ the *Herbrand expansion* of Π with respect to a certain Herbrand universe U. I will write $[\Pi]$ for the Herbrand expansion of Π with respect to the Herbrand universe U_Π of Π. Instead of $[\Pi]_{U_\mathcal{M}}$, where $U_\mathcal{M}$ denotes the Herbrand universe of some interpretation $\mathcal{M}$, I will simply write $[\Pi]_\mathcal{M}$. The Herbrand base of a program Π is denoted by At_Π, and its partial version is denoted by $\mathrm{Lit}_\Pi = \mathrm{At}_\Pi \cup \{\sim a : a \in \mathrm{At}_\Pi\}$.

Notice that a program clause may have a negative conclusion. This is essential for expressing explicit negative information or, logically speaking, constructible falsity. Already Sergot, Sadri, Kowalski et al. "did expect to need negative conclusions" [1986, p.379] in order to represent certain forms of negative knowledge in a legal reasoning system. They pointed out that standard Prolog lacks this capability.

---

[8]Notice that it was already suggested in [Lloyd & Topor 1984] to make Prolog more expressive by allowing general formulas instead of a conjunction of atoms in the premise of a program clause. Some authors, like Fitting [1986] and Miller [1989], adopt this syntactical generalization, while many others stick to the (unnecessarily) restricted notion of a program clause.

6.3   Model Theory

While the semantics of formulas is simply partial, the semantics of clauses and programs leads us one step towards constructive logic. Clauses can be viewed as first-degree implications in constructive logic. Because of this restriction to first-degree implication, however, the full apparatus of Kripke semantics for constructive logic as described e.g. in [Thomason 1969], [Rautenberg 1979], [Akama 1988a + 1990] is not needed.

A partial Herbrand interpretation[9] $\mathcal{M}$ is called a *model* of Π, symbolically $\mathcal{M} \models \Pi$, if for all $l \leftarrow F \in [\Pi]_{\mathcal{M}}$, $\mathcal{M} \models F$ implies $\mathcal{M} \models l$. Notice that this amounts to a non-contrapositive interpretation of $\leftarrow$, since a model of $\{\sim p(c), p(x) \leftarrow q(x)\}$ does not necessarily verify $\sim q(c)$.

Let $\mathrm{Mod}(\Pi) := \{M : \mathcal{M} \models \Pi\}$ and $\mathcal{M}_{\Pi} := \bigcap \mathrm{Mod}(\Pi)$.

Claim 6.1   *Every program Π has a least model, viz $\mathcal{M}_{\Pi}$.*

Proof: Obviously, if $\mathcal{M}_{\Pi}$ is a model of Π, it is the least one. Assume that it is not a model of Π. Then there is some ground clause $l \leftarrow F \in [\Pi]$ such that $\mathcal{M}_{\Pi} \models F$, but $l \notin \mathcal{M}_{\Pi}$. Since $\mathcal{M}_{\Pi}$ is the meet of all models of Π, there exists $M' \in \mathrm{Mod}(\Pi)$ with $l \notin M'$. But $\mathcal{M}' \models F$ (as a consequence of $\mathcal{M}' \geq \mathcal{M}_{\Pi}$ and the permanence principle). So, $\mathcal{M}' \models l$, i.e. $l \in M'$, which is a contradiction. $\square$

F is a *logical consequence* of Π, symbolically $\Pi \models F$, if every model of Π is also a model of F.

Claim 6.2 $\Pi \models F$ *iff* $\mathcal{M}_{\Pi} \models F$

This is a straightforward consequence of the previous claim and the permanence principle. It states that the model-theoretic consequence relation between a program and a formula is completely characterized by the least model of the program, and is the analogue of the well-known least Herbrand model characterization of positive logic programs.

6.4   Proof Theory

Constructive logic is the logic of direct proofs and refutations (also called 'direct logic' in [Kutschera 1985]). While there is only an indirect proof rule for negations (i.e. an indirect refutation rule) in Johannson's minimal and Heyting's intuitionistic logic, there is a general indirect proof rule in classical logic: *Reductio ad Absurdum*. None of these indirect rules is valid in constructive logic.

For the proof theory of constructive logic in general see e.g. [Prawitz 1965], [Almukdad & Nelson 1984], [Akama 1988b]. As an adaptation of it to logic programming a derivability relation between a program and a well-formed formula

---

[9]See 3.1.1.

is defined in the style of a natural deduction system by means of the introduction rules $(l),(\wedge),(\sim\wedge)$, $(\sim\sim)$ and (x).[10] Since we are interested in an operational proof theory which amounts to a procedural semantics we have to take care of the handling of autodependency loops. One way to do so is to exclude them by requiring the program to be in an appropriate sense wellfounded. Another way is to define a bottom-up procedure for constructing the hierarchy of derivable facts such that it is robust with respect to autodependency loops.

I now present the deduction rules for complex formulas (writing "$\Pi \vdash F,G$" as an abbreviation of "$\Pi \vdash F$ and $\Pi \vdash G$"):

$$(\wedge)(\sim\vee) \quad \frac{\Pi \vdash \sim F, \sim G}{\Pi \vdash \sim (F \vee G)}$$

$$(\vee) \quad \frac{\Pi \vdash F}{\Pi \vdash F \vee G} \qquad \frac{\Pi \vdash G}{\Pi \vdash F \vee G}$$

$$(\sim\wedge) \quad \frac{\Pi \vdash \sim F}{\Pi \vdash \sim (F \wedge G)} \qquad \frac{\Pi \vdash \sim G}{\Pi \vdash \sim (F \wedge G)}$$

$$(\sim\sim)(1) \quad \Pi \vdash 1$$

where F and G are ground formuas, and a non-ground formula is provable if some ground instance of it is,

$$(x) \quad \frac{\Pi \vdash F(t) \quad \text{for some ground term } t}{\Pi \vdash F(x)}$$

i.e. open formulas are interpreted as being implicitly existentially quantified.

In order to complete this definition of derivability relative to a program Π I have to say what it means for a ground literal to be derivable from Π. A straightforward way to define this is the following

$$(l)\text{iff} \quad \exists(l \leftarrow F) \in [\Pi] : \Pi \vdash F$$

However, this definition only works for 'well-behaved' programs which are called *wellfounded* according to the definition 6.1. In other cases it enters a loop (or some other form of an infinite descending chain). This problem does not arise in standard logic where the notion of derivability is not operational but simply requires the existence of a proof. It does not provide any effective method, though, to find the proof if it exists, or to find out that none exists, respectively. So, in the standard setting the following deduction rule would be added to the ones stated above:

$$\frac{\Pi \vdash F}{\Pi \vdash l} \quad \text{provided that } l \leftarrow F \in [\Pi]$$

---

[10] There is no need for elimination rules because $\vdash$ does not allow for arbitrary formulas in the premise.

A literal l, then, is derivable from Π if there is a derivation (i.e. a sequence of justified deduction-rule applications starting from '$\Pi \vdash 1$') for the sequent '$\Pi \vdash l$'.

Notice that there is no 'trivialization rule', so as to conclude anything from a contradiction. The principle *ex contradictione sequitur quodlibet*,

$$\frac{\Pi \vdash F, \sim F}{\Pi \vdash G}$$

which is fundamental in classical and intuitionistic logic, should only be assumed if one wants the effect of a single contradiction to be globally destructive, i.e. a single contradiction causes the loss of the entire information content of the database.

Alternatively, one might want to validate certain conclusions while invalidating others even in the presence of a contradiction. Of course, the admission of contradictions in the database gives rise to at least two questions: 1) What is the logical status of a contradictory piece of information? Can it count as true or false anyway? 2) How does a contradiction affect the manner of deriving further information possibly relying on the contradictory piece? For a short discussion of these questions and possible solutions see 6.9. Throughout this paper a liberal stance towards inconsistency is adopted, i.e. it is not required that programs are consistent[11] in order to be meaningful. Rather, the simultaneous derivability of a and $\sim a$ is considered as two pieces of information which are independent from each other, as in Belnap's system **B**, and in Nelson's system **N**.

6.4.1 Partial and Total Predicates

While a partial predicate in logic programming is best represented by two different clauses stating both its truth and its falsity conditions, it is sufficient for the representation of a total predicate to state either its truth or its falsity conditions. In order to handle total predicates a program has to define which predicates are positive-totally and which ones are negative-totally represented, for example by specifying two sets $\text{CWA}^+ = \{p, \ldots\}$ and $\text{CWA}^- = \{q, \ldots\}$ and applying them in the rule (l) which is extended by adding

$$(\text{CWA}) \qquad \frac{p \in \text{CWA}^+, \text{ and } \Pi \not\vdash p(t)}{\Pi \vdash \sim p(t)} \qquad \frac{q \in \text{CWA}^-, \text{ and } \Pi \not\vdash \sim q(t)}{\Pi \vdash q(t)}$$

where CWA refers to the Closed-World Assumption, and t stands for an arbitrary tuple of ground terms.

Since division in arithmetic is partial ("don't divide by zero!") there are natural cases of partial predicates even in mathematics, e.g.

$$p(x) \stackrel{def}{\Longleftrightarrow} \frac{1}{x-1} \geq 0 \quad (x \neq 1)$$

---

[11]Π is called consistent if there is no a such that $\Pi \vdash a, \sim a$.

which could be expressed by two rules stating both its truth and its falsity conditions:

$$p(x) \quad \leftarrow \quad x \neq 1 \wedge \frac{1}{x-1} \geq 0$$
$$\sim p(x) \quad \leftarrow \quad x \neq 1 \wedge \frac{1}{x-1} < 0$$

Obviously, $p(2)$ and $\sim p(0)$ hold, however, neither $p(1)$ nor $\sim p(1)$ holds.

I now illustrate the usefulness of the distinction between partially and totally represented predicates. The predicate *odd* which applies to odd numbers is exact and can be totally represented as in the following program (where s denotes the successor function of Peano arithmetic, i.e. $s(0) = 1$, $s(1) = 2$, etc.):

$$\Pi \quad = \quad \begin{cases} \text{CWA}^+ = \{odd, even\} \\ odd(s(0)) \\ odd(s(s(x))) \leftarrow odd(x) \\ even(x) \leftarrow \sim odd(x) \end{cases}$$

Since *odd* is positive-totally represented, we may infer its falsity according to the above CWA rule. So, for example, we obtain $\Pi \vdash even(s(s(0)))$ since $\Pi \nvdash odd(s(s(0)))$. Notice that without the CWA rule we would not obtain this conclusion.

While *odd* is analytical the predicate $on(x, y)$, which applies to two objects x and y when x is located on y, is empirical. In the case of an empirical predicate, even if it is exact, one has to check whether it is partially or totally represented in the database as the next example demonstrates:

$$\Pi \quad = \quad \begin{cases} on(a, b) \\ on(b, c) \\ above(x, y) \leftarrow on(x, y) \vee (on(x, z) \wedge above(z, y)) \\ \sim above(x, y) \leftarrow above(z, x) \wedge on(y, z) \end{cases}$$

It would be questionable to drop the falsity rule for *above* (the 4th clause of Π) and instead assume it to be totally represented by setting $\text{CWA}^+ = \{above, \ldots\}$. Because then we would get $\Pi \vdash \sim above(a, d)$, which is certainly not a piece of information implicitly encoded in Π – unless we assume that Π is a complete description of the 'world' in question. Rather, we should set $\text{CWA}^+ = \{\}$.

Notice that the query

$$above(a, x) \wedge \sim above(y, x)$$

succeeds with the answer $x = b$ and $y = c$. If negation-as-failure instead of strong negation had been used, the same query would flounder. As a matter of fact, there is no floundering query problem for logic programs with strong negation.

6.5 Wellfounded Programs

In the sequel I will make use of the following

Observation 6.1 $\Pi \vdash F$ *iff* $\exists K \in \mathrm{DNS}(F) \, \forall k \in K : \Pi \vdash k$

the proof of which is obtained by straightforward induction on F.

For a ground literal l, $\mathrm{Pre}^1(l)$ is defined as the set of its single-step literal predecessors, $\mathrm{Pre}^i(l)$ as the set of its ith-step literal predecessors, and $\mathrm{Pre}(l)$ as the set of all literals preceeding l in Π:

$$\mathrm{Pre}^1(l) \bigcup \{ K : K \in \mathrm{DNS}(F) \ \& \ l \leftarrow F \in [\Pi] \}$$

$$\mathrm{Pre}^{i+1}(l) \bigcup \{ \mathrm{Pre}^1(k) : k \in \mathrm{Pre}^i(l) \}$$

$$\mathrm{Pre}(l) \mathrm{Pre}^1(l) \ \cup \ \bigcup \{ \mathrm{Pre}(k) : k \in \mathrm{Pre}^1(l) \}$$

Intuitively speaking, $\mathrm{Pre}(l)$ collects all proper ground literals on which the derivability of l possibly depends.

Definition 6.1 (Wellfoundedness)   *A program* Π *is called* wellfounded,[12] *if*

$$\forall X \subseteq \mathrm{Lit}_\Pi \ \exists l \in X : \mathrm{Pre}(l) \cap X = \emptyset$$

For example, $\Pi = \{ q, p \leftarrow q, q \leftarrow p \}$ is not wellfounded since $\{p\} \cap \mathrm{Pre}(p) = \{p\} \neq \emptyset$. Also $\Pi = \{ p(s(0)), p(x) \leftarrow p(s(x)) \}$ is not wellfounded since $X = \{ p(s(0)), p(s(s(0))), \ldots \}$ does not contain any literal l such that $X \cap \mathrm{Pre}(l) = \emptyset$.

It is easy to see that in the case of a wellfounded program there is a stage $n < \omega$ such that for all $i \geq n$ we have that $\mathrm{Pre}^i(l) = \emptyset$. Thus, a ranking function d can be defined by setting $d(l) := \min\{ i : \mathrm{Pre}^i(l) = \emptyset \}$. If l is a fact, or if there is no rule for l, then $d(l) = 1$, otherwise $d(l) > 1$. As an immediate consequence of the definition of d we get the following

Observation 6.2   *Let* Π *be wellfounded, then*

$$\forall (l \leftarrow F) \in [\Pi] \, \forall K \in \mathrm{DNS}(F) \, \forall k \in K : d(k) < d(l).$$

Claim 6.3 (Adequacy)   *Let* Π *be wellfounded, then* $\Pi \vdash F$ *iff* $\Pi \models F$.

Proof: it suffices to show that if a formula is derivable from a program then the least model of the program verifies it, $\Pi \vdash F \Rightarrow \mathcal{M}_\Pi \models F$, and vice versa. Only the ground literal case, $F = l$, is treated. The complex cases follow by straightforward induction on the complexity of F.

---

[12] The notion of wellfoundedness corresponds to the notion of *acyclicity* of [Apt & Bezem 1990].

($\Rightarrow$) Let $d(l) = 1$, i.e. $l \leftarrow 1 \in [\Pi]$, and consequently, $\mathcal{M}_\Pi \models l$. Now let $d(l) = n$ and assume that the assertion holds for all literals k with $d(k) < n$. From $\Pi \vdash l$ we conclude that for some $l \leftarrow F \in [\Pi]$, $\Pi \vdash F$, or equivalently, $\exists K \in \mathrm{DNS}(F) \forall k \in K : \Pi \vdash k$. Since $d(k) < d(l)$ by observation 6.2, we may apply the induction hypothesis, hence $\mathcal{M}_\Pi \models k$, i.e. $K \subseteq M_\Pi$, and consequently, by observation 3.6, $\mathcal{M}_\Pi \models F$, implying that $\mathcal{M}_\Pi \models l$.

($\Leftarrow$) Conversely, we obtain $\Pi \vdash l$ from $\mathcal{M}_\Pi \models l$ by the same argument steps. $\square$

6.6 Non-Wellfounded Programs

6.6.1 Top-Down Derivation

The simplest case of a non-wellfounded program is $\Pi = \{p \leftarrow p\}$. Clearly, what is wanted is a decidable derivation procedure yielding $\Pi \not\vdash p$. In order to intercept such looping situations in the course of derivation an 'index' to the derivability relation is introduced.

Definition 6.2   *A program Π is called* semi-wellfounded *if for all $l \leftarrow F \in [\Pi]$,* Pre(l) *is finite.*

For an arbitrary set L of ground literals and a semi-wellfounded program Π, loop-tolerant derivability is defined as follows:

> (1) $\langle \Pi, L \rangle \vdash 1$
> (l)iff   for some $l \leftarrow F \in [\Pi]$ and some $K \in \mathrm{DNS}(F)$:
>   (i) $K \cap (L \cup \{l\}) = \emptyset$
>   (ii)   for all $k \in K : \langle \Pi, L \cup \{l\} \rangle \vdash k$

Notice that condition (i) provides a kind of loop-checking.[13]

Observation 6.3   *For a wellfounded program Π, we have $\Pi \vdash l$ iff $\langle \Pi, \emptyset \rangle \vdash l$.*

This is because wellfoundedness guarantees that condition (i) of the definition will be satisfied (proof by induction on $d(l)$).

Thus, derivability for semi-wellfounded programs Π can be defined by

> (l) $\Pi \vdash l \overset{def}{\Longleftrightarrow} \langle \Pi, \emptyset \rangle \vdash l$

Again, the derivability of complex and non-ground formulas is defined as in 6.4.

---

[13] For a detailled analysis of the general problem of loop-checking see e.g. [Apt, Bol & Klop 1989].

6.6.2 Bottom-Up Derivation

There is a quite natural notion of a single-step literal consequence operation C_Π^1 assigning to a set of literals X all literals which can be derived from it on the basis of Π in a single-step derivation:

$$C_\Pi^1(X) = \{l : l \leftarrow F \in [\Pi] \ \& \ X \vdash F\}$$

where X can be considered as a special case of a wellfounded program such that $X \vdash l$ if $l \in X$.

Observation 6.4 C_Π^1 is isotone, i.e. $X \subseteq Y \Rightarrow C_\Pi^1(X) \subseteq C_\Pi^1(Y)$.

We are especially interested in the properties of the sequence of sets of literals obtained through the iteration of C_Π^1:

$$C_\Pi^{i+1}(X) = C_\Pi^1(C_\Pi^i(X)) \qquad i \geq 1$$

Since C_Π^1 is isotone, $C_\Pi^i(\emptyset)$ is a monotonically increasing sequence constructing the least partial Herbrand model of Π in a similar way as in the fixed point semantics of positive logic programs by van Emden and Kowalski.[14]

$C_\Pi(\emptyset) = \bigcup_{i < \omega} C_\Pi^i(\emptyset)$ collects all literal consequences of $\emptyset$ with respect to Π, in other words, it is the literal theory determined by Π. At the same time, it also represents the least partial Herbrand model of Π.

Claim 6.4 $C_\Pi(\emptyset) = M_\Pi$

Proof: I first prove $M_\Pi \subseteq C_\Pi(\emptyset)$ by showing that $C_\Pi(\emptyset)$ is a model of Π. I abbreviate $C_\Pi^n(\emptyset)$ by Π^n and $C_\Pi(\emptyset)$ by Π^ω. For $l \leftarrow F \in [\Pi]$, if $\Pi^\omega \models F$, then, since Π^ω can be viewed as a wellfounded program, $\Pi^\omega \vdash F$, i.e. $\Pi^n \vdash F$ for some $n < \omega$. Thus, we obtain $l \in \Pi^{n+1}$, consequently, $\Pi^\omega \models l$.

Now, for the second part, $C_\Pi(\emptyset) \subseteq M_\Pi$, I show that $\Pi^n \subseteq M_\Pi$ for all $n \geq 1$ by induction on n. For $l \in \Pi^1$ we have $l \leftarrow 1 \in [\Pi]$. Since $M_\Pi \models 1$, we obtain $M_\Pi \models l$, i.e. $l \in M_\Pi$. For $l \in \Pi^n$ there is $l \leftarrow F \in [\Pi]$ with $\Pi^{n-1} \vdash F$, implying that for some $K \in \mathrm{DNS}(F)$ and all $k \in K$, $\Pi^{n-1} \vdash k$, i.e. $K \subseteq \Pi^{n-1}$. Hence, by the induction hypothesis, $K \subseteq M_\Pi$, implying $M_\Pi \models F$. As a consequence, $M_\Pi \models l$, i.e. $l \in M_\Pi$. $\square$

Claim 6.5 Let Π be semi-wellfounded and l be a proper ground literal, then $\Pi \vdash l$ iff $l \in C_\Pi(\emptyset)$.

Proof sketch: In order to prove the assertion by induction on the rank of l one has to define a modified ranking function for semi-wellfounded programs based on

---

[14] Cf. [Pearce & Wagner 1989].

an appropriately modified predecessor function. The argument, then, proceeds in the same way as for wellfounded programs.

As a corollary of the previous claims we obtain the adequacy of the loop-tolerant proof theory with respect to partial Herbrand semantics for semi-well-founded programs Π:

Claim 6.6   *Let Π be semi-wellfounded, then $\Pi \vdash F$ iff $\Pi \models F$.*

6.7   Logic Programs with Strong Negation as Vivid Rule Knowledge Bases

Logic programs with strong negation extend the framework of definite epistemic states by additionally allowing for conditional information expressed by program rules. Provided that for all clauses the variables occuring in the conclusion also occur in the premise (implying that there are only ground facts) they can also be viewed as rule knowledge bases of RV_0^+ where the facts of Π form the epistemic state X_Π, and the program rules form the set of rules R_Π. The system of logic programming with strong negation, thus, is a basic VKRS (in fact, it satisfies unrestricted reflexivity and monotonicity).

Definition 6.3 (Informational Ordering)   *Let Π and Π' be wellfounded propositional programs where all clauses are normalized, that is, premise formulas are conjunctions (resp. sets) of literals. Then, $F \subseteq$ Lit is defined to be more informative than $G \subseteq$ Lit (on the basis of Π) in the following way:*

$$F \geq_\Pi G \overset{def}{\Longleftrightarrow} \forall l \in G : l \in F, \text{ or } \exists (l \leftarrow F') \in \Pi : F \geq_\Pi F'$$

Finally, Π' is defined to be an informational extension of Π, if for any rule in Π, Π' contains one with the same conclusion but a less informative (i.e. less requiring) premise:

$$\Pi' \geq \Pi \overset{def}{\Longleftrightarrow} \forall (l \leftarrow F) \in \Pi \exists (l \leftarrow F') \in \Pi' : F \geq_{\Pi'} F'$$

Observation 6.5 $\geq_\Pi$ *and $\geq$ are preorderings.*

Claim 6.7 (Persistence)   *If $\Pi' \geq \Pi$ then $\Pi' \vdash F$ whenever $\Pi \vdash F$.*

The notions of answer and update equivalence can be used to compare programs:

Example 6.1   *Let $\Pi_1 = \{p \leftarrow \sim q, r\}$ and $\Pi_2 = \{\sim p \leftarrow q, r\}$. Then $\Pi_1 \simeq \Pi_2$, but $\Pi_1 \not\cong \Pi_2$.*

Example 6.2   *Let $\Pi_1 = \{p \leftarrow q, q\}$ and $\Pi_2 = \{p, q\}$. Then $\Pi_1 \cong \Pi_2$, and $\Pi_1 \leq \Pi_2$.*

Belnap's [1977] fixpoint semantics for information states applies also to logic programs with strong negation:

Claim 6.8 *A program clause $r \in R_\Pi$ is an ampliative and monotonic mapping from epistemic states to epistemic states.*

Proof: Any program clause $r = l \leftarrow F$ is a mapping $r : 2^{\text{Lit}} \to 2^{\text{Lit}}$ according to

$$r(X) = X \cup \{l\sigma : X \models F\sigma\}$$

where σ ranges over ground substitutions for F. That r is ampliative, $X \subseteq r(X)$, follows immediately from the definition. Suppose $X_1 \subseteq X_2$. By persistence, $X_2 \models F\sigma$ whenever $X_1 \models F\sigma$, implying that

$$r(X_1) = X_1 \cup \{l\sigma : X_1 \models F\sigma\} \subseteq X_2 \cup \{l\sigma : X_2 \models F\sigma\} = r(X_2). \quad \square$$

Claim 6.9 *For any program Π, $\langle X_\Pi, R_\Pi \rangle$ is an information state in the sense of Belnap. Consequently, the fixpoint $R_\Pi(X_\Pi)$ collects all ground literals inferrable from Π.*

Thus, we obtain a fixpoint semantics for logic programs with strong negation,

$$\Pi \vdash F \quad \text{iff} \quad R_\Pi(X_\Pi) \models F$$

Notice that the requirement of monotonicity of rules depending on the persistence of premise formulas is crucial for the construction of the fixpoint $R_\Pi(X_\Pi)$. In a realistic setting, however, where negation-as-failure would be allowed in premise formulas these would be no longer persistent, hence rules would be ampliative but nonmonotonic mappings from epistemic states to epistemic states. One then loses the elegant lattice-theoretical methods suggested by Belnap which are all based on continous (i.e. monotonic) mappings.

6.8 Relation to Other Logics

6.8.1 Reducibility to Positive Logic

Strong negation can be eliminated from logic programs in the same way as it can be eliminated from formulas of constructive logic (see 3.3.2). For this purpose, a reductive translation of an arbitrary formula F to a positive formula F^* not containing any occurence of $\sim$ is defined:

$$(F \wedge G)^* F^* \wedge G^*$$
$$(F \vee G)^* F^* \vee G^*$$
$$(\sim(F \wedge G))^* (\sim F)^* \vee (\sim G)^*$$

$$(\sim(F \vee G))^* (\sim F)^* \wedge (\sim G)^*$$
$$(\sim\sim F)^* F^*$$
$$(\sim a)^* \text{not\_}a$$
$$a^* a$$
$$1^* 1$$
$$\Pi^* \{l^* \leftarrow F^* : l \leftarrow F \in \Pi\}$$

where not_a is an ad-hoc introduced new atom.

Claim 6.10 $\Pi \vdash F \Rightarrow \Pi^* \vdash F^*$

Proof: I only prove that the assertion holds for $F = l$, the other cases follow by straightforward induction on F. It is shown by induction on n that for arbitrary $n, l \in \Pi^n \Rightarrow l^* \in (\Pi^*)^n$.

Let $n = 1$ and $l \in \Pi^1$. For $l = a$ we have $a \leftarrow 1 \in \Pi$, consequently $a \leftarrow 1 \in \Pi^*$ which implies $a \in (\Pi^*)^1$. For $l = \sim a$ we obtain in the same way that not_$a \in (\Pi^*)^1$. So, in any case, $l^* \in (\Pi^*)^1$.

Now, consider $l \in \Pi^n$, i.e. $\exists (l \leftarrow G) \in [\Pi]$ such that $\Pi^{n-1} \vdash G$. We proceed by induction on G:

Case 1 : $G = k$ Since $k \in \Pi^{n-1}$ the n-induction hypothesis can be applied, hence $k^* \in (\Pi^*)^{n-1}$. By the translation procedure, $l^* \leftarrow k^* \in \Pi^*$, consequently $l^* \in (\Pi^*)^n$.

Case 2 : $G = \sim(H_1 \wedge H_2)$ From $\Pi \vdash \sim(H_1 \wedge H_2)$ it can be concluded that $\Pi \vdash \sim H_1$ or $\Pi \vdash \sim H_2$, so, by the G-induction hypothesis, $\Pi^* \vdash (\sim H_1)^*$ or $\Pi^* \vdash (\sim H_2)^*$, yielding $\Pi^* \vdash (\sim(H_1 \wedge H_2))^*$. As a consequence, we get $\Pi^* \vdash l^*$.

Cases 3,4 : $G = \sim\sim H, G = H_1 \wedge H_2$ Straightforward. $\square$

A reduction of the above kind seems to characterize strong negation. No analogous result holds for the negation of classical, intuitionistic and minimal logic, which define a stronger semantical link between a formula and its negation. In [Pearce 1992] this property of strong negation is studied in more detail.

As announced in the introduction, it is possible to reduce logic programs with strong negation to definite Horn clause programs. For this purpose the *normal form* $\hat{\Pi}$ of a program Π is defined as $\hat{\Pi} = \{l \leftarrow \bigwedge K : l \leftarrow F \in \Pi \ \& \ K \in \text{DNS}(F)\}$. The transformation of a program Π, then, is achieved in two steps: normalization and positivization (the order of which does not matter) resulting in the definite Horn clause program $\hat{\Pi}^*$.

Observation 6.6 $\Pi \vdash l \Rightarrow \hat{\Pi}^* \vdash l^*$

Proof: In view of claim 6.10 it suffices to show that $\hat{\Pi}$ is equivalent to Π, in other words, $l \in \Pi^i$ iff $l \in \hat{\Pi}^i$ for all $i < \omega$. Let $i = 1$, i.e. $l \in \Pi^1$, consequently $l \leftarrow 1 \in [\Pi]$. Since $l \leftarrow 1 \in [\hat{\Pi}]$, it also holds $\hat{\Pi} \vdash l$. Now let $i = n$, i.e.

$\exists (l \leftarrow F) \in [\Pi]$ such that $\Pi^{n-1} \vdash F$, which is equivalent to $\exists K \in \mathrm{DNS}(F)$ such that $K \subseteq \Pi^{n-1}$. So, by the induction hypothesis, $K \subseteq \hat{\Pi}^{n-1}$, and since $l \leftarrow \bigwedge K \in [\hat{\Pi}]$ by the definition of $\hat{\Pi}$, we obtain $\hat{\Pi} \vdash \bigwedge K$, consequently $\hat{\Pi} \vdash l$.
□

6.8.2 Relation to Classical Logic

Conditional derivability can be defined as a ternary relation between a program, a premise formula and a conclusion formula (both of which are ground), as follows:

$$\Pi, F \vdash G \stackrel{def}{\Longleftrightarrow} \forall K \in \mathrm{DNS}(F): \Pi \cup K \vdash G$$

Conditional derivability captures the notion of positive implication between (implication-free) formulas in the sense of the metalogical implication $\Pi \vdash F \Rightarrow \Pi \vdash G$. Indeed, as a consequence of transitivity (see observations 5.7 and 5.12), one obtains the *cut rule*, also called Lemma Redundancy here:

$$\Pi \vdash F \ \& \ \Pi, F \vdash G \ \Rightarrow \ \Pi \vdash G$$

The following theorem characterizes the difference between derivability from a consistent program in constructive and in classical logic: the latter one can be 'imitated' by means of the former one by adding all instances of the *tertium non datur* to a program.

Claim 6.11 *Let Π be a consistent program with strong negation. A ground formula G is derivable from*

$$P = \bigwedge \{\sim F \vee l : l \leftarrow F \in [\Pi]\}$$

in classical logic, $P \vdash_{cl} G$, iff

$$\Pi, \bigwedge \{a \vee \sim a : a \in A\} \vdash G$$

where A is the Herbrand base of $\{\Pi, G\}$.

Proof: I show that $\sim P \vee G$ is a classical tautology *iff* for all $K \in \mathrm{DNS}(\bigwedge \{a \vee \sim a : a \in A\})$ such that $\Pi \cup K$ is consistent, $\Pi \cup K \models G$, where A is the Herbrand base of $\{P, G\}$. I start with the following

Observation 6.7 *Let A be the Herbrand base of some ground formula F and let $\mathcal{M}$ decide A, i.e. $\forall a \in A : a \in M^+$ or $a \in M^-$. Then,*
(i) if F is a classical tautology, $\mathcal{M} \models F$
(ii) $\mathcal{M} \models \sim F$ iff $\mathcal{M} \not\models F$

Now, let $\sim P \vee G$ be a classical tautology and let A be the Herbrand base of $\{P, G\}$. If $\mathcal{M} \models \Pi \cup K$, $\mathcal{M}$ decides A. Consequently, according to observation (i), $\mathcal{M} \models \sim P \vee G$, which implies by (ii) that $\mathcal{M} \not\models P$ or $\mathcal{M} \models G$. Since $\mathcal{M} \not\models P$ contradicts our assumption that $\mathcal{M} \models \Pi \cup K$, we obtain $\mathcal{M} \models G$.

Conversely, let $\Pi \cup K \models G$ for K as specified above, and let $\mathcal{M}$ be total.

Case 1 : $\mathcal{M} \models \Pi \cup K$ Then, by our assumption, $\mathcal{M} \models G$.

Case 2 : $\mathcal{M} \not\models \Pi \cup K$ We have to distinguish between two subcases:

a) $\mathcal{M} \not\models \Pi$ Then there is $l \leftarrow F \in [\Pi]_{\mathcal{M}}$ such that $\mathcal{M} \models F$ and $\mathcal{M} \not\models l$, which is equivalent to $\mathcal{M} \not\models P$, and hence to $\mathcal{M} \models \sim P$ by observation (ii).

b) $\mathcal{M} \models \Pi$ Let $K' = M \cap (A \cup \{\sim a : a \in A\})$. Then $K' \in \text{DNS}(\bigwedge\{a \vee \sim a : a \in A\})$ and $\Pi \cup K'$ is consistent, consequently, $\Pi \cup K' \models G$, implying $\mathcal{M} \models G$.

Thus, in any case, either $\mathcal{M} \models G$, or $\mathcal{M} \models \sim P$, which implies that $\mathcal{M} \models \sim P \vee G$. $\square$

A similar result for his logic of inexact predicates was proved by Cleave [1974].

6.8.3   Relation to Forward Chaining

Herre [1991] has defined a *calculus of forward chaining*, **FC**, for program formulas

$$l_1 \wedge l_2 \wedge \ldots \wedge l_n \rightarrow l_0$$

by means of four rules (substitution, $\wedge$-introduction, $\exists$-introduction, and detachment) together with contraction and permutation. Forward chaining is a frequently used technique in rule-based expert systems.

It turns out that **FC** is a slightly less general variant of logic programming with strong negation. It is constructively correct with respect to classical logic, and it is complete with respect to 3-valued logic. For more details the reader is referred to [Herre & Pearce 1992].

6.8.4   Relation to Constructive Logic

Although there is no implication connective in the object language, program clauses can be viewed as first-degree intuitionistic implications. Thus, it is not surprising that there is a close relationship between logic programs with strong negation and the paraconsistent constructive logic **N**.

Claim 6.12   *Let* $P := \{F \rightarrow l : l \leftarrow F \in [\Pi]\}$, *and* $G \in L(\sim, \wedge, \vee)$. *Then*

$$\Pi \vdash G \quad \textit{iff} \quad P \vdash_N G$$

Proof sketch: ($\Rightarrow$) It can be easily shown by induction on n that $\Pi^n \vdash G$ implies $P \vdash_N G$. Let $\Pi^0 \vdash G$, i.e. $G = 1$, and trivially, $P \vdash_N 1$. Now, let $\Pi^n \vdash G$, i.e. $\exists K \in \mathrm{DNS}(G) : K \subseteq \Pi^n$, and consequently $\forall k \in K \exists (k \leftarrow F) \in [\Pi] : \Pi^{n-1} \vdash F$, i.e. $\exists L \in \mathrm{DNS}(F) : L \subseteq \Pi^{n-1}$. It follows by the induction hypothesis that $\forall l \in L : P \vdash_N l$, implying that $P \vdash_N F$. Hence also $P \vdash k$ for all $k \in K$, and consequently $P \vdash_N G$.

($\Leftarrow$) Proof by induction on the length of a derivation. Since P contains only implications a literal can only be inferred from P in $\mathbf{N}$ by means of the elimination rule for $\rightarrow$.[15] The induction base step is again trivial. Assume that $P \vdash_N^i G$ implies $\Pi \vdash G$ for all $i < n$. Let $P \vdash_N^n G$, i.e. $\exists K \in \mathrm{DNS}(G) \forall k \in K : P \vdash_N^n k$, implying that $\exists (F \rightarrow k) \in P : P \vdash_N^i F$ such that $i < n$. By the induction hypothesis, then, $\Pi \vdash F$. Since $k \leftarrow F \in [\Pi]$ for all $k \in K$, this implies $\Pi \vdash k$, and consequently, $\Pi \vdash G$. $\square$

Harrop formulas with implication-free premises in constructive logic are equivalent to logic programs: their disjunctive normal set is a singleton consisting of clauses and facts, that is, a logic program. Therefore, logic programs with strong negation represent a special class of constructive formulas, namely those formulas describing an AND-OR-NON-graph, that is, a directed graph with four kinds of links where certain links are AND-grouped. AND-OR-NON-graphs are graphical representations of normalized Harrop formulas with implication-free premises. In such a graph a node represents an atom and a link represents the rule connective '$\leftarrow$' together with one of the four inference patterns: true if true ($\longleftarrow$), false if true ($\longleftarrow$), true if false ($\longleftarrow$), and false if false ($\longleftarrow$).

For instance, the normalization of

$$F = \sim (p \rightarrow q \vee r) \wedge (\sim q \rightarrow (r \rightarrow \sim p \wedge s))$$

yields $\mathrm{DNS}(F) = \{P\}$ where

$$P = \begin{cases} 1 \rightarrow p,\ 1 \rightarrow \sim q,\ 1 \rightarrow \sim r \\ \sim q \wedge r \rightarrow \sim p \\ \sim q \wedge r \rightarrow s \end{cases}$$

which is graphically represented in figure 6.1.

In the propositional case it is straightforward to generalize the notion of a program clause, such that every formula of constructive logic can be represented by a set of *Harrop programs*. I will only sketch this generalization here. A Harrop program consists of *Harrop clauses* $l \leftarrow F$ where $F \in L(1, \sim, \wedge, \vee, \rightarrow)$. The derivability relation between a program and a formula is extended by defining

$$\Pi \vdash F \rightarrow G \forall \Sigma \in \mathrm{DNS}(F) : \Pi \cup \Sigma \vdash G$$
$$\Pi \vdash \sim (F \rightarrow G) \Pi \vdash F \wedge \sim G$$

---

[15] In other logics, such as classical and intuitionistic, where indirect rules, such as (Contraposition), (Negation as Inconsistency), or even (Reductio ad Absurdum), are valid this would not be true: one would have more possibilities to infer a literal from a set of implications, and would therefore in general get more consequences than can be inferred from the program !

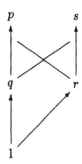

Figure 6.1: The AND-OR-NON-graph of P.

These generalized clauses allow for nested implications in the premise. A disjunctive normal set in constructive logic, then, can be viewed as a set of such programs. Consequently, every formula in constructive logic is equivalent to a certain set of Harrop programs:

$$F \vdash_N G \quad \text{iff} \quad \forall \Pi \in \text{DNS}(F) : \Pi \vdash G$$

The possibility to represent any formula of constructive logic as a set of logic programs emphasizes

1. the significance of logic programs as the definite constituents of general information, and

2. the adequacy of constructive logic as the semantical basis for logic programming.

6.9 Inconsistency Handling

There are at least three strategies for dealing with inconsistency which seem to make sense as alternative options in logic programming with strong negation.[16] This, from a pragmatically point of view, useful plurality is rendered possible by the concept of strong negation which does not commit one to a particular way of handling contradictions.

In mathematical logic (classical as well as intuitionistic), the occurence of a contradiction is usually considered a devastating event causing the immediate loss of all meaning of the theory in question – it becomes *trivial* according to

---

[16] Of course, from the viewpoint of pure logic, there are far more possibilities to distinguish between different notions of inconsistency in formal systems, see e.g. the literature on paraconsistent logics.

the principle *ex contradictione sequitur quodlibet (ECSQ)*. We have the same possibility for logic programs with strong negation: we could postulate that as soon as a contradiction is derivable from a program, every formula is derivable from it. We would then have models of a pogram only if it is consistent, but there would be no model for it, if it is not.

In terms of a logical semantics for information processing it seems to be inadequate and over-restrictive to consider a single contradiction as being globally destructive. Rather one should either not bother at all about contradictions, or, one should maintain their effect locally, i.e. consider the specific pieces of information which are contradictory as useless, and regard all remaining information not depending on it as unaffected. In [Wagner 1991c] the former approach is called *liberal* and the latter one *conservative*, resp. *skeptical*. Both approaches lead to an extended partial semantics where every program, even an inconsistent one, has a model.

The liberal approach leads to the 4-valued logic **B** of Belnap. In this framework one could add *negation-as-inconsistency*,[17] $\neg$, by defining

$$(\neg) \quad \frac{\Pi \vdash F \rightarrow a \wedge \sim a \quad \text{for some atom } a}{\Pi \vdash \neg F}$$

If (ECSQ) would hold, one could more simply define $\neg F := F \rightarrow \sim F$. Of course, before that, intuitionistic implication has to be added by defining

$$(\rightarrow) \quad \frac{\Pi, F \vdash G}{\Pi \vdash F \rightarrow G}$$

The conservative approach leads to a nonmonotonic logic which apparently has not yet been studied in the literature. In [Wagner 1991c] the principles of this logic are described, and its relation to Reiter's logic of normal defaults is discussed.

I illustrate the different ways of inconsistency handling with an example. The information that Peter is married either to Susan or Jill, Tom is married to Jill, and a man is not married to a woman if she is married to another man, is encoded in the following way,

$$\Pi_1 \quad = \quad \left\{ \begin{array}{r} m(P,S) \sim m(P,J) \\ m(P,J) \sim m(P,S) \\ m(T,J) \\ \sim m(x,y) m(z,y) \wedge z \neq x \end{array} \right.$$

$\Pi_2 = \Pi_1 \cup \{\sim m(T,J)\}$ is inconsistent. In the liberal approach we would get the contradictory conclusions $\Pi_2 \vdash m(T,J)$ and $\Pi_2 \vdash \sim m(T,J)$. While conclusions based on contradictory information, like $\Pi_2 \vdash m(P,S)$, remain valid, we do not obtain any trivial conclusions, like $\Pi_2 \vdash m(T,P)$, with the help of (ECSQ).

---

[17] Negation-as-inconsistency in the absence of (ECSQ) is well-known from Johannson's *minimal logic*.

In conservative reasoning we not only give up the trivialization of Π_2 but we also discard contradictory information, so $\Pi_2 \nvdash m(T, J)$, consequently giving up all conclusions based on contradictory information, so for instance $\Pi_2 \nvdash m(P, S)$.

Notice that with strong negation we could not conclude from Π_1 that Peter is not married to Tom, which we could do by means of negation-as-inconsistency, $\Pi_1 \nvdash \sim m(P, T)$ but $\Pi_1 \vdash \neg m(P, T)$.

6.10   Related Work

In [Blair & Subrahmanian 1989] a paraconsistent logic programming language allowing for the processing of undetermined and overdetermined information is presented. In this language atoms have so-called annotations, i.e. they have an explicit truth value attribute. Thus, for an atom a, there are the four expressions, $a : f$, $a : \perp$, $a : \top$ and $a : t$ standing for 'a is false', 'a is undetermined', 'a is overdetermined' and 'a is true', respectively. Clauses of such *paraconsistent programs* have the form

$$l_0 : \mu_0 \leftarrow l_1 : \mu_1, \ldots, l_n : \mu_n$$

where l_i is a literal and μ_i is its annotation, i.e. one of the four truth-values.

I will now sketch a translation from a 'paraconsistent program' without undetermined premises, i.e. atoms annotated by $\perp$, to a program with strong negation.

1. Replace all negative literals $\neg a : \mu$ by $a : \bar{\mu}$ where $\bar{f} = t$, $\bar{\perp} = \perp$, $\bar{\top} = \top$ and $\bar{t} = f$ everywhere in the program.

2. Replace everywhere in the program $a : t$ by a and $a : f$ by $\sim a$.

3. Replace all clauses $a : \top \leftarrow X$ by the two clauses $a \leftarrow X$ and $\sim a \leftarrow X$.

4. Delete all clauses $a : \perp \leftarrow X$.

5. Lastly, replace everywhere in the program the annotated atoms which remained in the body of clauses, $a : \top$ by $a \wedge \sim a$.

The translated program, then, has the same set of consequences as the paraconsistent version. It seems clear, however, that the language of logic programs with strong negation is much more natural, and that it is more elegant to express the undeterminacy and the overdeterminacy of some piece of information by the usual logical expressions (containing weak negation[18] if necessary).

---

[18] An undetermined literal, $l : \perp$, can be expressed by means of strong negation in combination with *weak negation* – as the expression $-l \wedge -\sim l$.

Gelfond and Lifschitz [1990] have proposed "Logic Programs with Classical Negation". They consider program clauses of the form

$$l_0 \leftarrow l_1, \ldots, l_m, -l_{m+1}, \ldots, -l_n$$

where '$-$' denotes negation-as-failure (interpreted autoepistemically).

I will restrict the comparison here to the monotonic fragment of 'logic programs with classical negation', i.e. programs not containing negation-as-failure. This can be done without loss of generality since the remarks to be made concern only the supposed 'classicality' of the second negation of Gelfond and Lifschitz which is in no way related to negation-as-failure.

Gelfond and Lifschitz define the semantics of a program in terms of answer sets. The *answer set* of a program Π (without negation-as-failure), denoted by $\alpha(\Pi)$, is defined as the smallest subset $S \subseteq \text{Lit}_\Pi$, such that

(AS1) For any rule $l \leftarrow F \in [\Pi]$, if $S \models F$, then $l \in S$.

(AS2) If for some $a \in \text{At}_\Pi$, both $a \in S$, and $\sim a \in S$, then $S = \text{Lit}$.

Notice that because of (AS2) the answer set semantics is explosive. If (AS2) is dropped, however, the resulting definition of an answer set, say $\alpha'(\Pi)$, agrees with the least partial Herbrand model $\mathcal{M}_\Pi$, i.e. $\alpha'(\Pi) = \mathcal{M}_\Pi$.

What is named "classical negation" by Gelfond and Lifschitz is, in fact, strong negation, since it does not satisfy the essential property of classical negation, the principle *tertium non datur* (which, in turn, is related to the semantic principle of bivalence constituting classical logic). This is easy to see. Consider $\Pi = \{p \leftarrow q, \ p \leftarrow \sim q\}$. Π has the unique answer set $\alpha(\Pi) = \emptyset$ (corresponding to the semantics presented here, where $\mathcal{M}_\Pi = \emptyset$). In other words, p is not a consequence of Π, which clearly violates the semantics of classical logic where the principle of Reasoning by Cases,

$$\frac{X, F \vdash G \qquad X, \sim F \vdash G}{X \vdash G}$$

yields the conclusion p, since $\Pi, q \vdash p$, as well as $\Pi, \sim q \vdash p$.

So, it is not surprising that Gelfond and Lifschitz arrive, yet in a less general form, at the same results:

1. They express the CWA in terms of their "classical" negation in the same way as is done here in terms of strong negation in 6.4.1.

2. They also demonstrate the eliminability of their "classical" negation as described for strong negation in 6.8.1.

6.11   Conclusion

Strong negation is probably the most appropriate candidate of logically well-defined negations to be implemented in logic programming. It solves the frequently discussed problem of representing and processing explicit negative information. Other researchers, notably Gelfond and Lifschitz [1990+1991], though apparently not aware of the logical concept, have also realized its usefulness and feasibility.

I believe that future systems of logic programming will incorporate two or even three kinds of negation: negation-as-failure, strong negation and possibly negation-as-inconsistency. This will greatly enhance the capability of a logic program as a means of knowledge representation.

Chapter 7

Vivid Reasoning on the Basis of Rules

Generalizing the vivid reasoning approach of Levesque it is proposed that the basic expressions of knowledge representation are (positive and negative) facts and rules with two kinds of negation. For the processing of explicit negative information the concept of strong negation already known from partial logic and constructive logic is introduced. It is shown how the interaction between strong negation and weak negation (alias negation-as-failure) in the premise of a rule can be modeled within the framework of partial logic. The expressive power of this extended language allows to handle various phenomena of commonsense reasoning, and thus complies with the basic principles of vividness: cognitive adequacy and computational feasibility. As shown in the appendix, it can be implemented as a Prolog meta-interpreter in a straightforward manner.

In order to deal with the operational problem of loops the notions of a *wellfounded* and a *weakly wellfounded* KB, corresponding to 'acyclic' and 'locally stratified' programs, are introduced and a loop-tolerant recursive proof-theory for such KBs is defined. It is shown that the induced natural model of a weakly wellfounded KB is *perfect*.

7.1 Introduction

In contrast to Levesque's original proposal the language of a vivid knowledge base (VKB) will not be restricted to simple facts and definite clauses here. Rather, the aim is a far more expressive framework for vivid reasoning where the basic expressions of knowledge representation are *conditional facts*, i.e. inference rules with literal conclusions. The proposed generalizations can also be applied to database systems, like those of a relational and a deductive database (RDB and

DDB). Table 7.1 illustrates the suggested evolution of knowledge bases where successively weak and strong negation are introduced to systems which have been purely positive in the beginning.

RDB: $\{a\} \vdash G, \quad G \in L(\wedge, \vee)$	
DDB: $\{a \leftarrow F\} \vdash G, \quad F, G \in L(\wedge, \vee)$	
RDB + weak negation: $\{a\} \vdash G, \quad G \in L(\wedge, \vee, -)$	RDB + strong negation: $\{l\} \vdash G, \quad G \in L(\wedge, \vee, \sim)$
DDB + weak negation: $\{a \leftarrow F\} \vdash G$ $F, G \in L(\wedge, \vee, -)$	DDB + strong negation: $\{l \leftarrow F\} \vdash G$ $F, G \in L(\wedge, \vee, \sim)$
RDB + weak + strong negation: $\{l\} \vdash G, \quad G \in L(\wedge, \vee, -, \sim)$	
DDB + weak + strong negation: $\{l \leftarrow F\} \vdash G, \quad F, G \in L(\wedge, \vee, -, \sim)$	

Table 7.1: Knowledge Base Evolution Level 1: only positive information, Level 2: either implicit or explicit negative information, Level 3: both explicit and implicit negative information.

Inference rules with literal conclusions have, besides their bottom-up interpretation as mappings between KBs, another interpretation, namely as procedures which are evaluated in a top-down fashion. Since the aim is to establish an operational proof theory which amounts to a procedural semantics, it has to be taken care of the handling of dependency loops. One way to do so is to exclude them by requiring the VKB to be in an appropriate sense wellfounded. Another way is to use loop-tolerant procedures, e.g. bottom-up procedures constructing the hierarchy of derivable facts. However, for a certain class of VKBs which are called *non-weakly-wellfounded* the proof theory presented is not complete. These VKBs represent indefinite information requiring major modifications in order to be handled adequately (see also the discussion in 7.8).

7.2   Vivid Knowledge Bases

A vivid knowledge base V consists of inference rules, or *clauses*, of the form $l \leftarrow F$ (read "l if F") where l is a proper literal and F an arbitrary formula. Such rules are also called *conditional facts*. A rule with premise 1 is called a *fact*, and $l \leftarrow 1$ is often abbreviated by l. Examples of conditional facts are

$$\sim flies(x) \leftarrow emu(x) \vee penguin(x)$$

$$switch\_on\_light \leftarrow dark \wedge -illuminated$$

Since the semantics of quantifiers is of no concern here, only ground queries to VKBs where all variables occuring in the premise of a rule also occur in the conclusion will be considered. This restriction guarantees that we only deal with ground formulas in the course of derivations, and therefore stay, essentially, on a propositional level (avoiding the problem of 'floundering queries').

Strictly speaking, (conditional) vivid facts are ground, i.e. (conditional) assertions about specific individuals. So, if l is not ground it represents something like a 'general fact' which might be viewed as a collection of the resp. ground instances, or alternatively, as an assertion about arbitrary objects[1]. Consequently, a vivid knowledge base V containing non-ground conditional facts will be viewed as a dynamic representation of the corresponding set of ground conditional facts formed by means of the current domain of individuals U and denoted by $[V]_U$. Formally,

$$[V]_U = \{l\sigma \leftarrow F\sigma \mid l \leftarrow F \in V \text{ and } \sigma : \text{Var}(l, F) \rightarrow U\}$$

where σ ranges over all mappings from the set of variables of l and F into the set of all constant symbols U. σ is called a *ground substitution* for $l \leftarrow F$ and $[V]_U$ the *Herbrand expansion* of V with respect to a certain (finite) Herbrand universe U. The Herbrand expansion of V with respect to the Herbrand universe U_V induced by V, i.e. the set of all constant symbols occuring in V, is denoted by $[V]$. Instead of $[V]_{U(\mathcal{M})}$, where $U(\mathcal{M})$ is the Herbrand universe of some model $\mathcal{M}$, we will simply write $[V]_{\mathcal{M}}$.

The language of VKBs is designed for expressing four kinds of information:

definite positive information (as represented in RDBs) by means of atoms, like, for instance, the positive facts

$$likes(Peter, Tom),\ hates(Tom, Peter)$$

definite negative information by means of strongly negated atoms, like the negative fact that Peter dislikes Mary,

$$\sim likes(\ Peter,\ Mary)$$

conditional information by means of conditional facts, like

$$\sim likes(x, y) \leftarrow hates(x, y),$$

stating that x dislikes y whenever x hates y

implicit negative information by means of weak negation (in queries and the premise of conditional facts), as in the conditional statement that two persons can be placed next to each other, if it is not the case that one of them dislikes the other,

$$placeNextToEachOther(x, y) \leftarrow -(\sim likes(x, y) \lor \sim likes(y, x))$$

---

[1] Cf. [Fine 1985].

The intuitive reading of $\sim p$ is 'p is falsifiable' or 'p is known to be false' whereas $-p$ would mean 'p is not verifiable' or 'p is not known to be true'. Likewise, $-\sim p$ can be read as 'p is not falsifiable' or 'p is not known to be false' which obviously doesn't reduce to p (not disliking another person does not amount to liking her !).

Although the intention of this framework appears to be the handling of definite information, by the combination of conditionals and weak negation indefiniteness has crept into it. This problem is further discussed in 7.8.

A propositional (or monadic) vivid knowledge base can be viewed as a logical representation of an AND-OR-NOT-NON-graph, that is, an AND-OR-graph with eight different kinds of links representing the inference patterns:

true if true	$\longleftarrow$
true if false	$\longleftarrow$
true if not true	$\longleftarrow$
true if not false	$\longleftarrow$
false if true	$\longleftarrow$
false if false	$\longleftarrow$
false if not true	$\longleftarrow$
false if not false	$\longleftarrow$

Table 7.2: Inference patterns and corresponding link types of an AND-OR-NOT-NON-graph.

The following monadic VKB

$$V = \left\{ \begin{array}{l} p(c),\ p(d),\ \sim q(d) \\ \sim r(x) \leftarrow -\sim q(x) \wedge p(x) \\ r(x) \leftarrow -p(x) \vee \sim q(x) \end{array} \right.$$

is graphically represented in figure 7.1.

7.3   Model Theory

Let $\mathcal{M}$ be a partial Herbrand model as defined in 3.1.1. $\mathcal{M}$ is a *model* of a vivid knowledge base V, symbolically $\mathcal{M} \models V$, if for all $l \leftarrow F \in [V]_{\mathcal{M}}$, $\mathcal{M} \models l$ whenever $\mathcal{M} \models F$. Notice that this interpretation of the conditional $\leftarrow$ is non-contrapositive, i.e. a model of $\{p \leftarrow q, \sim p\}$ does not necessarily support $\sim q$ since the model may leave q either undetermined, or overdetermined.

For example, abbreviating the predicate names 'bird' and 'flies' by b and f, and the individual names 'Molly' and 'Tweety' by the constant symbols m and

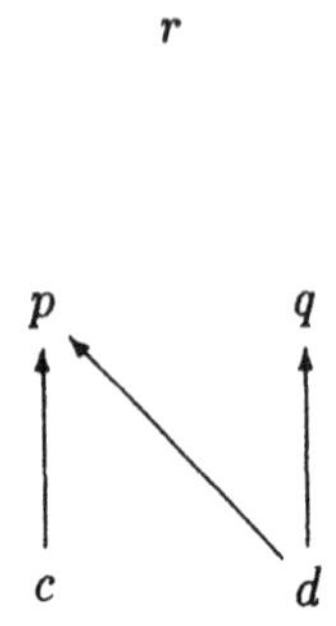

r

Figure 7.1: The AND-OR-NOT-NON-graph of V.

t, respectively, we obtain

$$\langle \{b(t), b(m), f(t), \}, \{f(m)\}\rangle \quad \models$$
$$\{f(x) \leftarrow b(x) \wedge -\sim f(x), \sim f(m), b(t), b(m)\}$$

Claim 7.1 *Every vivid knowledge base V without weak negation has a least model, viz the meet of all its models.*

Proof: V corresponds to a logic program with strong negation. Therefore, the assertion holds according to claim 6.1. $\square$

An arbitrary VKB, however, need not have a least model. Consider, for example,

$$V_1 = \left\{ \begin{array}{l} p \leftarrow -\sim q \\ \sim q \leftarrow -p \end{array} \right.$$

which has two minimal models, $\{p\}$ and $\{\sim q\}$, the meet of which, i.e. the empty set, is no longer a model.

What is an appropriate notion of model-theoretic consequence in this setting ? Obviously, the standard notion ('a consequence is what is valid in all models') does not do the job. For instance, $-p$ should be a logical consequence of the empty VKB, $V = \emptyset$. But, of course, there are models of V, where $-p$ is not valid, e.g. $M = \{p\}$. So it seems that a notion of *preferred* model is needed in order to capture the information content of a VKB model-theoretically.

Are the preferred models of a VKB simply its minimal ones ? Consider $V_2 = \{\sim q \leftarrow -p\}$. In minimal model semantics, V_2 would be equivalent to V_1 above, both sharing the same minimal models, viz. $\{p\}$ and $\{\sim q\}$. Yet, V_2 has not the same information content as V_1, it actually represents more information (by fewer rules !), since $\sim q$ is an intended consequence of V_2 but not of V_1.

Therefore, the minimal model semantics is not adequate (in the above example it is not complete with respect to the intended consequence relation). Certain examples indicate that it is not clear whether the intended consequences should include all 'minimal consequences'. In 7.8, I will argue that neither the perfect nor the wellfounded models of a VKB are, in general, the preferred models looked for.

7.4   Wellfounded VKBs

Concerning the recursive structure of a VKB, the most straightforward way to define derivability for ground literals is the following

$$(l)\text{if}\quad \exists(l \leftarrow F) \in [V] : V \vdash F$$
$$(-l)\text{if}\quad \forall(l \leftarrow F) \in [V] : V \vdash -F$$

Notice that the right hand side of $(-l)$ is decidable because for a VKB V without function symbols, the Herbrand expansion $[V]$ is finite. However, this definition only works for 'well-behaved' VKBs which are called *wellfounded* according to the definition below. Otherwise it enters a loop. This problem does not arise in standard logic where the notion of derivability is not operational but simply requires the existence of a proof. It does not provide any effective method, though, to find the proof if it exists, or to find out that none exists, respectively. So, in the standard setting we would add the following deduction rules to the ones stated above:

$$\frac{V \vdash F \quad \text{for some } l \leftarrow F \in [V]}{V \vdash l} \qquad \frac{V \vdash -F \quad \text{for all } l \leftarrow F \in [V]}{V \vdash -l}$$

A formula F, then, is derivable from V if there is a derivation (i.e. a sequence of justified deduction rule applications starting from '$V \vdash 1$') for the sequent '$V \vdash F$'.

In order to say what it means for a VKB to be wellfounded a few definitions are needed. For a literal l we define $\mathrm{Pre}_V^1(l)$, the set of its single-step literal predecessors, $\mathrm{Pre}_V^i(l)$, the set of its ith-step literal predecessors, and $\mathrm{Pre}_V(l)$, the set of all literals preceding it in V (in the sequel, the index V will be omitted):

$$\mathrm{Pre}^1(l)\bigcup\{\, E^+ \cup E^- : E \in \mathrm{DNS}(F) \ \& \ l \leftarrow F \in [V]\,\}$$
$$\mathrm{Pre}^{i+1}(l)\bigcup\{\, \mathrm{Pre}^1(k) : k \in \mathrm{Pre}^i(l)\,\}$$
$$\mathrm{Pre}(l)\mathrm{Pre}^1(l) \ \cup \ \bigcup\{\mathrm{Pre}(k) : k \in \mathrm{Pre}^1(l)\}$$

Intuitively speaking, $\mathrm{Pre}(l)$ collects all proper literals on which the derivability of l possibly depends. V is called *wellfounded*, if for every $l \leftarrow F \in [V]$ we have

$l \notin \mathrm{Pre}(l)$, i.e. there are no loops in $[V]$ – neither positive nor negative ones.[2]

Example 7.1 *For instance, if $V_3 = \{q \leftarrow p \wedge -(\sim q \wedge -r), \sim q \leftarrow s, p \leftarrow 1\}$ then $\mathrm{Pre}^1(q) = \{p, \sim q, r\}$, $\mathrm{Pre}^2(q) = \{s\}$, $\mathrm{Pre}^3(q) = \emptyset$ and $\mathrm{Pre}(q) = \{p, \sim q, r, s\}$.*

In the case of a wellfounded VKB V one can define a ranking function $d = d_V$ (the index V will be omitted). Let $d(l) = \min\{i : \mathrm{Pre}^i(l) = \emptyset\}$ and $d(V) = \max\{d(l) : l \leftarrow F \in [V]\}$. If l is a fact, or if there is no rule for l in V, then $d(l) = 1$, otherwise $d(l) > 1$. The definition of d can be extended to arbitrary formulas by setting $d(-l) = d(l)$, and $d(F) = \max\{d(e) : e \in \bigcup \mathrm{DNS}(F)\}$.

As an immediate consequence of the definition of d we get the following

Observation 7.1 *Let V be wellfounded, then*

$$\forall (l \leftarrow F) \in [V] \, \forall E \in \mathrm{DNS}(F) \, \forall e \in E : \ d(e) < d(l).$$

Every wellfounded VKB has a natural stratification of its Herbrand expansion induced by d:

$$V(0)\emptyset$$

$$V(i)\{l \leftarrow F \in [V] : d(l) = i\} \qquad \text{for } i \geq 1$$

$$V^j \bigcup_{k \leq j} V(k) \qquad \text{for } j, k \geq 0$$

Observation 7.2 *Let V be wellfounded. Then $V \vdash -F$ iff $V \not\vdash F$.*

Proof by two-fold induction on $d(V)$ and on F.

Observation 7.3

1. $V \vdash F$ *iff* $\exists E \in \mathrm{DNS}(F) \, \forall e \in E : V \vdash e$

2. $V \vdash -F$ *iff* $\forall E \in \mathrm{DNS}(F) \, \exists e \in E : V \vdash \bar{e}$

Proof by straightforward induction on F.

The natural stratification of a wellfounded VKB V leads to an inductive construction of its natural model $\mathcal{M}_V$:

$$\mathcal{M}_V(0)\emptyset$$

$$\mathcal{M}_V(i)\{l : l \leftarrow F \in V(i) \ \& \ \mathcal{M}_V^{i-1} \models F\} \qquad \text{for } i \geq 1$$

$$\mathcal{M}_V^j \bigcup_{k \leq j} \mathcal{M}_V(k) \qquad \text{for } j, k \geq 0$$

$$\mathcal{M}_V \, \mathcal{M}_V^{d(V)}$$

---

[2]In fact, the notion of wellfoundedness corresponds to the notion of *acyclicity* of [Apt & Bezem 1990] in the following sense: if $a \leftarrow b_1, \ldots, b_n, not \ c_1, \ldots, not \ c_m$ is a clause of an acyclic program (without function symbols) then $a \leftarrow b_1 \wedge \ldots \wedge b_n \wedge -c_1 \wedge \ldots \wedge -c_m$ is a clause of the corresponding wellfounded VKB.

Example 7.1 (continued)   *We get $d(V) = d(q) = 3$ and the following strat-ification:*

$$V(1)\{p \leftarrow 1\}$$
$$V(2)\{\sim q \leftarrow s\}$$
$$V(3)\{q \leftarrow p \wedge -(\sim q \wedge -r)\}$$

$$M_V(1) = \{p\}, \quad M_V(2) = \{\}, \quad M_V(3) = \{q\}$$

Finally, $M_V = M_V^3 = \{p, q\}$.

Observation 7.4 M_V *is supported, i.e.* $\forall l \in M_V \; \exists l \leftarrow F \in [V]$ *such that* $M_V \models F$.

Proof: If $l \in M_V$ then for some n, $l \in M_V(n)$, and consequently there is $l \leftarrow F \in V(n)$, i.e. $d(l) = n$, and $M_V^{n-1} \models F$, implying that for some $K \in \mathrm{DNS}(F)$, $K^+ \subseteq M_V^{n-1} \subseteq M_V$, and $K^- \cap M_V^{n-1} = \emptyset$. It remains to show that $K^- \cap M_V = \emptyset$ in order to obtain $M_V \models F$. Since for all $k \in K^-$, $d(k) < d(l) = n$, and on the other hand, for all $k \in M_V - M_V^{n-1}$, $d(k) \geq n$, it follows that $K^- \cap (M_V - M_V^{n-1}) = \emptyset$, and hence $K^- \cap M_V = \emptyset$. $\square$

Claim 7.2 $M_V \models V$

Proof by induction on $d(V)$: If $d(V) = 1$, then V contains only facts and the assertion is obvious. Now assume that for some V, $d(V) = n$, and consequently, $M_V = M_V^n$. It has to be shown that if $l \leftarrow F \in [V]$ and $M_V \models F$, then $l \in M_V$. Without loss of generality assume that $d(l) = d(V) = n$, implying that $l \leftarrow F \in V(n)$. From $M_V \models F$ it follows that there is some $E \in \mathrm{DNS}(F)$ such that $E^+ \subseteq M_V^n$ and $E^- \cap M_V^n = \emptyset$. It suffices now to show that $E^+ \subseteq M_V^{n-1}$, implying $M_V^{n-1} \models F$, which by the induction hypothesis yields $l \in M_V^{n-1} \subseteq M_V^n$. Since M_V^n is supported, it holds that for all $k \in E^+ \subseteq M_V^n$, there is a rule $k \leftarrow G \in V^n$ such that $M_V^n \models G$. Since $k \in M_V(d(k))$ and $d(k) < d(l) = n$, it follows that $M_V(d(k)) \subseteq M_V^{n-1}$, implying that $k \in M_V^{n-1}$, and hence $E^+ \subseteq M_V^{n-1}$. $\square$

Observation 7.5   *Let $e \in \mathrm{XLit}$, then $V \vdash e \Rightarrow \forall i \geq d(e) : V^i \vdash e$.*

Proof by induction on $d(e)$: Let $d(e) = 1$. If $e = l$ then $V \vdash l$ implies that $l \leftarrow 1 \in V(1)$, and therefore trivially $V^i \vdash l$ for all $i \geq 1$. If on the other hand $e = -l$ then $V \vdash -l$ implies that for all $l \leftarrow F \in V(1)$, $V \vdash -F$. Since $V(1)$ contains only rules with premise 1 this condition is only satisfied if there is no rule for l in $V(1) \subseteq [V]$. Consequently, $V^i \vdash -l$ for all $i \geq 1$.

Now, let $d(e) = n$. If $e = l$ then $V \vdash l$ implies that there is a rule $l \leftarrow F \in V(n)$ such that for some $E \in \mathrm{DNS}(F)$ and for all $f \in E$, $V \vdash f$. Since $d(f) < d(e) = n$, we may apply the induction hypothesis, hence, for all $f \in E$

and for all $i \geq d(f)$, $V^i \vdash f$, and consequently, also $V^i \vdash F$ for all $i \geq d(F)$, implying $V^i \vdash l$ for all $i \geq n$.

If $d(e) = n$ and $e = -l$ then $V \vdash -l$ implies that for all $l \leftarrow F \in [V]$, $V \vdash -F$, i.e. $\forall E \in \mathrm{DNS}(F) \exists f \in E : V \vdash -f$. Since $d(-f) = d(\bar{f}) = d(f) < d(e) = n$, we may apply the induction hypothesis, hence, $V^i \vdash \bar{f}$ for all $i \geq d(f)$. As a consequence, for all $l \leftarrow F \in [V]$ and for all $i \geq d(F)$, $V^i \vdash -F$, implying that $V^i \vdash -l$ for all $i \geq d(l)$. $\square$

Claim 7.3 $\mathcal{M}_V$ *is adequate:* $\mathcal{M}_V \models F$ *iff* $V \vdash F$.

Proof: It suffices to show, by induction on $d(l)$, that

(i) $V \vdash l \Rightarrow l \in \mathcal{M}_V$

(ii) $V \nvdash l \Rightarrow l \notin \mathcal{M}_V$

since from (i) and the contraposition of (ii) follows that

$$l \in \mathcal{M}_V \quad \text{iff} \quad V \vdash l$$

and finally, by $V \vdash -l \Leftrightarrow V \nvdash l \Leftrightarrow l \notin \mathcal{M}_V \Leftrightarrow \mathcal{M}_V \models -l$, also

$$\mathcal{M}_V \models e \quad \text{iff} \quad V \vdash e$$

Now, let $d(l) = 0$. Then,

(i) $V \vdash l \Rightarrow l \leftarrow 1 \in [V] \Rightarrow l \in \mathcal{M}_V$, since $\mathcal{M}_V$ is a model.

(ii) $V \nvdash l$ implies that there is no rule $l \leftarrow F \in [V]$, and consequently $l \notin \mathcal{M}_V$, since $\mathcal{M}_V$ is supported.

Suppose that the assertion has been proven for all l with $d(l) < n$, and let $d(l) = n$. Then,

(i) $V \vdash l \Rightarrow \exists l \leftarrow F \in V(n) \exists E \in \mathrm{DNS}(F) \forall e \in E : V \vdash e$. Since $d(e) < d(l)$, by the induction hypothesis, $\mathcal{M}_V \models e$, i.e. $E^+ \subseteq \mathcal{M}_V$ and $E^- \cap \mathcal{M}_V = \emptyset$, implying that $\mathcal{M}_V \models F$. Since $\mathcal{M}_V$ is a model of V this yields $l \in \mathcal{M}_V$.

(ii) $V \nvdash l \Rightarrow \forall l \leftarrow F \in [V] : V \vdash -F$, i.e. $\forall E \in \mathrm{DNS}(F) \exists e \in E : V \vdash \bar{e}$. Since $d(\bar{e}) < d(l)$, by the induction hypothesis, $\mathcal{M}_V \models \bar{e}$, i.e. if $e = l$ then $l \notin \mathcal{M}_V$, and if $e = -l$ then $l \in \mathcal{M}_V$, and consequently in any case, for all $E \in \mathrm{DNS}(F)$, either $E^+ \not\subseteq \mathcal{M}_V$, or $E^- \cap \mathcal{M}_V \neq \emptyset$, implying that $\mathcal{M}_V \not\models F$. Since $\mathcal{M}_V$ is supported, $l \notin \mathcal{M}_V$. $\square$

There are non-wellfounded VKBs, however, which seem to have unique well-determined logical consequences. For example, $V_1 = \{q, p \leftarrow q, q \leftarrow p\}$ and $V_2 = \{p \leftarrow -p\}$ are not wellfounded since in both cases $p \in \mathrm{Pre}(p)$. Nevertheless, it should be possible to derive p from V_1. Thus, a more general and more robust (viz positive-loop-tolerant) way of derivation is needed.

7.5　Weakly Wellfounded VKBs

The weak predecessors of a literal, $\text{WPre}(l)$, are defined as follows:

$$X(l) \bigcup \{E : E \in \text{DNS}(F) \ \& \ l \leftarrow F \in [V]\}$$

$$\text{WPre}^1(l)X^-(l) \cup \bigcup \{\text{Pre}(k) : k \in X^-(l)\}$$

$$\text{WPre}(l)\text{WPre}^1(l) \cup \bigcup \{\text{WPre}(k) : k \in X^+(l)\}$$

$\text{WPre}(l)$ collects all weakly negated predecessor literals of l together with their resp. predecessors.

V is called *weakly wellfounded*, if for every $l \leftarrow F \in [V]$, $l \notin \text{WPre}(l)$, i.e. only positive but no negative loops are allowed.

Observation 7.6 *If $a \leftarrow b_1, \ldots, b_n, notc_1, \ldots, notc_m$ is a clause of a locally stratified program[3] (without function symbols) then the corresponding VKB containing the clause $a \leftarrow b_1 \wedge \ldots \wedge b_n \wedge -c_1 \wedge \ldots \wedge -c_m$ is weakly wellfounded.*

Weakly wellfounded VKBs also have a natural stratification induced by a ranking function $d'(l)$ which differs from $d(l)$ by cutting off positive loops in $\text{Pre}(l)$ in the same way as in the loop-tolerant proof procedure defined below. This stratification also induces a 'natural' model $\mathcal{M}_V$.

7.5.1　The Induced Model $\mathcal{M}_V$ is Perfect

Przymusinski [1988] has introduced the notion of a *perfect* model capturing the minimization priorities between the predicates of a logic program with respect to weak falsity. I will give similar definitions for VKBs.

Definition 7.1　*A model $\mathcal{M}'$ is called* preferable to $\mathcal{M}$ *(with respect to a given VKB), symbolically $\mathcal{M}' \preceq \mathcal{M}$, if $\forall l \in \mathcal{M}' - \mathcal{M} \ \exists k \in \mathcal{M} - \mathcal{M}' : k \in \text{WPre}(l)$. $\mathcal{M}$ is called* perfect *if there are no models preferable to it.*

Claim 7.4　*A weakly wellfounded VKB V has a unique perfect model, namely $\mathcal{M}_V$.*

Proof sketch: Let $\mathcal{M}$ be an arbitray model of V, and let $M^i = M \cap \{l : l \leftarrow F \in V^i\}$. It can be shown by induction on i that $\mathcal{M}_V^i \preceq \mathcal{M}^i$.

The concept of perfect models, however, is not the solution to the problem of finding a proper definition of the preferred models of an arbitrary VKB. Rather it points to one important property those models must have.

---

[3] Cf. [Przymusinski 1988].

7.5.2 Loop-Tolerant Top-Down Derivation

The simplest case of a non-wellfounded (but nevertheless weakly wellfounded) VKB is $V = \{p \leftarrow p\}$. Clearly, a derivation procedure should yield $V \nvdash p$. In order to intercept such looping situations in the course of derivation an 'index' to the derivability relation is introduced. In the following definition $L \subseteq \text{Lit}$ contains all literals on which the derivation at a given stage must not depend:

$$
\begin{array}{ll}
(1) & \langle V, L \rangle \vdash 1 \\
(l)\text{if} & \exists(l \leftarrow F) \in [V] \, \exists E \in \text{DNS}(F): \\
& \text{(i)} \quad E^+ \cap (L \cup \{l\}) = \emptyset, \text{ and} \\
& \text{(ii)} \quad \forall e \in E : \langle V, L \cup \{l\} \rangle \vdash e \\
(-l)\text{if} & \forall(l \leftarrow F) \in [V] \, \forall E \in \text{DNS}(F): \\
& \text{(i)} \quad E^+ \cap (L \cup \{l\}) \neq \emptyset, \text{ or} \\
& \text{(ii)} \quad \exists e \in E : \langle V, L \cup \{l\} \rangle \vdash \bar{e}
\end{array}
$$

Notice that condition (i) provides a kind of loop-checking for positive loops.

Observation 7.7 *For a wellfounded VKB V, we have $V \vdash l$ iff $\langle V, \emptyset \rangle \vdash l$.*

This is because wellfoundedness guarantees that condition (i) of the definition will be satisfied. Thus, derivability for (not necessarily wellfounded but at least) weakly wellfounded VKBs can be defined by

$$
(l) \quad V \vdash l \quad \text{iff} \quad \langle V, \emptyset \rangle \vdash l
$$

With respect to loop-tolerant derivability the induced model of a weakly wellfounded VKB is adequate:

Claim 7.5 *For any weakly wellfounded VKB V, and any ground formula F,*

$$
\mathcal{M}_V \models F \quad \text{iff} \quad V \vdash F
$$

7.5.3 Bottom-Up Derivation

There is a quite natural notion of a single-step literal consequence operation C_V^1 assigning to a set of literals $X \subseteq \text{Lit}$ all literals which can be derived from it on the basis of V in a single-step derivation:

$$
C_V^1(X) = \{l : l \leftarrow F \in [V] \ \& \ X \vdash F\}
$$

Here, X can be considered as a wellfounded VKB consisting of simple facts, so $X \vdash l$ iff $l \in X$. However, this operator is, in general, not monotone. We are especially interested in the properties of the sequence of sets of literals obtained through the iteration of C_V^1:

$$
C_V^{i+1}(X) = C_V^1(C_V^i(X))
$$

If V does not contain weak negation, i.e. the language of V is restricted to $\langle \wedge, \vee, \sim, 1 \rangle$, $C_V^i(\emptyset)$ is a monotonically increasing sequence constructing the least partial Herbrand model of V in a similar way as in the fixed point semantics of positive logic programs by van Emden and Kowalski.

The question now is: what are the restrictions on V such that at some stage m of the construction the sequence becomes stationary,

$$C_V^n(\emptyset) = C_V^m(\emptyset)$$

for all $n \geq m$, in other words, a fixed point of C_V^1 is obtained.

Claim 7.6   *The operator C_V^1 has a fixed point if V is weakly wellfounded.*

This fixed point is denoted by $C_V(\emptyset)$ since it collects all literal consequences of $\emptyset$ with respect to V, in other words, it is the literal theory determined by V. At the same time, it also represents the least partial Herbrand model of V.

Claim 7.7   *Let V be weakly wellfounded, then $V \vdash l$ iff $l \in C_V(\emptyset)$.*

7.6   VKBs as Rule Knowledge Bases

A VKB can be viewed as a rule knowledge base, and hence also as an information state in the sense of Belnap. As in the case of logic programs with strong negation, the set of all facts of V represents a definite epistemic state, and the rules of V, collected in R_V, represent ampliative mappings between definite epistemic states. Thus, the VKRS of VKBs corresponds to the rule-based extension of fact bases, RV_0.

However, rules in RV_0 are not monotonic, in general, since they may have non-persistent premise formulas if weak negation is involved. Therefore, the composition of such rules is not necessarily an upper bound for them, and hence, in the general case there is no 'supremum rule' subsuming all $r \in R_V$. In the case of a (weakly) wellfounded VKB V any composition of rules

$$r_n \circ r_{n-1} \circ \ldots \circ r_1$$

respecting the induced ordering of V,

$$d(\mathrm{Concl}(r_i)) \leq d(\mathrm{Concl}(r_{i+1})) \qquad \text{for } i = 1, \ldots, n-1$$

can be regarded as a representative of the intensional meaning R_V of V, and the extensional meaning of V is then given by $R_V(X_V)$ where X_V collects the facts of V.

7.7 Related Work

The idea of logic programming with two kinds of negation seems to become a new paradigm in knowledge representation and AI. Following [Gelfond & Lifschitz 1990], such programs are called *extended*. In their standard syntax they consist of clauses of the form

$$l_0 \leftarrow l_1, \ldots, l_m, -l_{m+1}, \ldots, -l_n$$

where the premise is a conjunction of extended literals.

In some approaches to extended programs the concept of an *extended interpretation*

$$\mathcal{I} = \langle I^t, I^{dt}, I^f, I^{df} \rangle$$

is used. It does not only desginate the true and false atoms I^t and I^f, but in addition also those atoms which are false or true by default, I^{df} and I^{dt}.[4] Even if one wants to allow for paraconsistency, that is, $I^t \cap I^f \neq \emptyset$, it is essential for such interpretations to require

(Default Consistency) $I^{df} \cap I^t = \emptyset$ & $I^{dt} \cap I^f = \emptyset$

7.7.1 The 'Answer Set Semantics' of Gelfond and Lifschitz

Gelfond and Lifschitz [1990] propose 'Logic Programs with Classical Negation'. Normal logic programs are extended by adding another negation, which they call 'classical' but which, in fact, is the strong negation known from the systems **B** and **N** (see 6.10). Gelfond and Lifschitz define a kind of autoepistemic semantics for '$-$'. For this purpose they introduce the reductive transformation Π^S of an extended program Π with respect to a 'belief set' $S \subseteq$ Lit as the program obtained from $[\Pi]$ by:

1. discarding each rule which contains a weakly negated literal $-l$ in the premise such that $l \in S$,

2. deleting all weakly negated literals in the premises of the remaining rules.

Recall the definition of an answer set $\alpha(\Pi)$ of a program Π without weak negation (in 6.10): essentially, a minimal set of literals closed under all program rules. An *answer set* of an extended program Π is then any set $S \subseteq$ Lit such that

$$S = \alpha(\Pi^S)$$

---

[4]Notice that for models of definite programs it is not necessary to designate I^{df} and I^{dt} because they are simply the complements of I^t and I^f: $I^{df} = \text{At}_P - I^t$, and $I^{dt} = \text{At}_P - I^f$.

A literal is defined to follow from a program Π if it is in all answer sets of Π.

The answer set semantics for extended programs is a straightforward generalization of the *stable model* semantics of [Gelfond & Lifschitz 1988] for normal programs. It is attractive because it assigns a meaning to every program. It conservatively extends the interpretation of (weakly) wellfounded programs. And it seems to capture the indefiniteness of programs, like e.g. $\{p \leftarrow -q, q \leftarrow -p\}$, for which it yields two answer sets $\{p\}$ and $\{q\}$, and hence the indefinite conclusion $p \vee q$.

However, there are simple examples (e.g. where the clause $p \leftarrow -p$ is involved) for which the results of the stable model semantics do not seem to be intuitive, or are at least debatable (see the discussion of the barber paradox in the next section). Another criticism is that there is no efficient operational proof theory for stable model semantics: one has to make a guess first, and then check if it is correct (by testing its stability). Finally, inconsistency handling in answer set semantics is not vivid: either the answer set explodes according to ECSQ embodied in (AS2), by the definition of α, or inconsistency leads to non-intuitive results when no answer set is obtained, as demonstrated by the following example:

Example 7.2 (Gelfond & Lifschitz 1990)   *The following program*

$$\Pi = \{p \leftarrow -\sim p, q \leftarrow p, \sim q \leftarrow p\}$$

has no answer set. Therefore, trivially, any literal follows from it.

7.7.2 The 'Wellfounded Semantics with Explicit Negation' of Pereira and Alferes

The approach of Pereira and Alferes [1992], called $\mathcal{WFSX}$, also generalizes the stable model semantics to extendend programs. Pereira and Alferes consider the intuitive requirement that $\sim l$ should imply $-l$, called *Coherence*, as essential for $\mathcal{WFSX}$.

A $\mathcal{WFSX}$ interpretation is an extended interpretation $\langle I^t, I^{dt}, I^f, I^{df} \rangle$ satisfying

(Coherence) $I^t \subseteq I^{dt}$ & $I^f \subseteq I^{df}$

implying that $I^t \cap I^f = \emptyset$.

The diagram of a $\mathcal{WFSX}$ interpretation $\mathcal{I}$ is the corresponding set of extended literals $I = \{e \in \text{XLit} : \mathcal{I} \models e\}$, where

$$\mathcal{I} \models aa \in I^t$$
$$\mathcal{I} \models \sim aa \in I^f$$
$$\mathcal{I} \models -aa \in I^{df}$$
$$\mathcal{I} \models -\sim aa \in I^{dt}$$

In order to avoid notational redundancy, a $\mathcal{WFSX}$ interpretation will also be represented as a pair of true plus true-per-default and false plus false-per-default atoms:

$$\langle\, I^t + (I^{dt} - I^t),\ I^f + (I^{df} - I^f)\,\rangle$$

For an interpretation $\mathcal{I}$, Π^I denotes the transformed program obtained from $[\Pi]$ by:

1. removing all rules containing $-l$ in the premise such that $l \in I$,

2. removing all rules containing l in the premise such that $\tilde{l} \in I$,

3. deleting all $-l$ in premises if $-l \in I$,

4. replacing all remaining weakly negated literals in premises by the logical constant **u**.

This transformation yields a program without weak negation which has a least partial Herbrand model $\mathcal{M}_{PI}$.

Any proper partial Herbrand interpretation (or consistent set of literals) $J \in \mathrm{Cons}(2^{\mathrm{Lit}})$ can be extended to a $\mathcal{WFSX}$ interpretation $Coh(J)$ by setting

$$Coh(J) = J \cup \{-\tilde{l} : l \in J\} \cup \overline{\mathrm{Lit}_J - J}$$

Finally, a $\mathcal{WFSX}$ interpretation $\mathcal{I}$ is called an *extended stable model* of an extended program Π if

$$Coh(M_{PI}) = I$$

The $\mathcal{WFSX}$ semantics, then, defines a literal to follow from a program Π if it is in all extended stable models of Π.[5]

Since the semantics of Pereira and Alferes is based on the stable model approach it inherits the problems and weaknesses of it mentioned in the previous section. On the other hand, the more general notion of a model and the built-in coherence principle seem to be attractive features which can be useful in further improvements of the semantics with respect to inconsistency handling. Such an improvement is proposed in [Pereira, Alferes and Aparicio 1992].

7.7.3 Other Work on the Use of Two Kinds of Negation

In [Sakama 1992] an 'extended wellfounded semantics for paraconsistent logic programs' with two kinds of negation based on extended models is proposed. As opposed to the approach of Pereira and Alferes, models are neither required

---

[5] Since an inconsistent program does not have any extended stable model, everything follows from it, i.e. the $\mathcal{WFSX}$ semantics is explosive.

to be coherent nor consistent (Sakama even ommits the condition of Default Consistency although he seems to assume it implicitly). Sakama motivates his notion of an extended model by relating it to Ginsberg's bilattice for default logic [Ginsberg 1986]. Recall that for the diagram I of an extended model $\mathcal{I}$, and an extended literal $e \in$ XLit, $I \vdash e$ iff $e \in I$. Sakama presents a generalization of Przymusinski's [1989] fixpoint construction of the wellfounded model of a program Π based on two single-step inference operators P_1^+ and P_1^- taking an interpretation $I \subseteq$ XLit and a set of tentatively derived, resp. failed, literals $X \subseteq$ Lit, and providing additional tentatively derived, resp. failed, literals:

$$P_1^+(I,X)\{l \mid \exists l \leftarrow F \in [\Pi] : I \cup X \vdash F\}$$
$$P_1^-(I,X)\{l \mid \forall l \leftarrow F \in [\Pi] : I \cup \overline{X} \vdash -F\}$$

Keeping I fixed these operators are iterated in the following way:

$$P_{n+1}^*(I,X) P_1^*(I, P_n^*(I,X)) \qquad \text{for } * = +, - \text{ and } n \geq 1$$
$$P^+(I,X) \bigcup_{n<\omega} P_n^+(I,X)$$
$$P^-(I,X) \bigcap_{n<\omega} P_n^-(I,X)$$

Finally, the wellfounded model of Π is obtained by a two-fold iteration process:

$$M_P^0 \emptyset$$
$$M_P^{n+1} M_P^n \cup P^+(M_P^n, \emptyset) \cup \overline{P^-(M_P^n, \text{Lit}_P)}$$
$$M_P \bigcup_{n<\omega} M_P^n$$

Sakama also presents a refinement of his P^+ and P^- operators in order to take into account the fact that some conclusions of an extended program may depend on contradictory premises, and this should be recorded in some way. However, he does not relate the inconsistency of a literal l to its failure, i.e. to the validity of $-l$ (as it is done e.g. in [Wagner 1991c]), which seems to be a shortcoming.

Also Przymusinski [1990] considers logic programs with two kinds of negation. But while he gives a three-valued semantics to the weak negation, he treats the strong negation in a non-logical fashion which seems to be questionable, and makes it impossible to relate both negations, or to introduce 'smart' forms of inconsistency handling.

Kowalski and Sadri [1990] use the stable model semantics as the basis for their treatment of extended logic programs. Guided by principles of abductive reasoning they incorporate a preference of negative over positive rules which does not permit to represent and process explicit negative information in a general way. Negative rules can only be used to express exceptions to their positive counterparts.

Extending earlier work by Sandewall a 3-valued approach to nonmonotonic reasoning based on weak and strong negation is proposed in [Doherty 1990]. In this approach, however, rules are represented by material implication which does not seem to be fully adequate, since 3-valued material implication does not satisfy the deduction theorem – an essential for a genuine conditional operator.

Da Costa et al. [1990] develop a resolution calculus for logics with weak and strong negation based on the principle of paraconsistency. They argue that their approach can be applied to the problem of conflicting expert knowledge in an expert system. Since vivid logic is also paraconsistent, the main difference seems to be that in the approach of da Costa et al. general disjunctive clauses are the basic expressions whereas vivid logic handles clauses as rules capable of representing genuine conditional information.

7.8 Weak Negation and Indefinite Knowledge

Consider the following simplified version of the well-known Barber Paradaox:

$$(b-) \qquad shave(b, x) \leftarrow -shave(x, x)$$

This rule says that b shaves x whenever there is no explicit information on x shaving himself. Suppose V contains $(b-)$ and the fact $shave(c, c)$. Then we can conclude that b shaves a but not that b shaves c. Does the barber b shave himself ?

According to the *wellfounded semantics* introduced in [Van Gelder, Ross & Schlipf 1988] he does not. According to any semantics based on partial models he does. The argument for this is quite simple: Let $V = \{s(b, x) \leftarrow -s(x, x), s(c, c)\}$ and let $\mathcal{M}$ be a partial Herbrand model of V. Suppose M^+ does not contain $s(b, b)$, then $\mathcal{M}$ satisfies the premise of $s(b, b) \leftarrow -s(b, b) \in [V]$. Consequently, it contains $s(b, b)$. So, every partial Herbrand model, i.e. also every – in whatever sense – preferred model, of V necessarily supports the fact that b shaves himself. This argument can only be rejected if one denies the adequacy of partial models for such non-wellfounded KBs.

The meaning of weak negation in vivid logic (inspired by negation-as-failure) seems to be well-defined only for weakly wellfounded VKBs. For such a VKB exists a unique maximally preferred model induced by its implicit stratification. This induced model provides an adequate account of the information content of a VKB. Notice, however, that it comes with a strong syntactic flavor, and, in this respect, departs from the traditional concept of a syntax-independent model-theoretic semantics.

The question now is: can the semantics of weakly wellfounded VKBs be conservatively extended to the general case of arbitrary VKBs, and then yield intuitively correct answers to all generic examples of non-wellfounded KBs ? Any proposal related to the semantics of general, or extended, logic pograms or deductive databases, like the perfect models, the stable models or the wellfounded semantics, is subject to this requirement.

Non-weakly-wellfounded VKBs represent indefinite, i.e. disjunctive, knowledge. I have not defined derivability for them. But it seems that e.g. $p \leftarrow -p$ stands for $p \vee p$ which by the idempotence of disjunction is equivalent to p. While the stable models semantics cannot handle such clauses at all (they are unsatisfiable), the wellfounded semantics is too skeptical: it does not allow to conclude anything from them.

Another example seems to be similarly generic with respect to the indefiniteness of non-weakly-wellfounded VKBs: $V = \{p \leftarrow -q, q \leftarrow -p\}$. There is no least (and also no perfect) model for V but we obtain two minimal models: $\{p\}$ and $\{q\}$. Hence, $p \vee q$ is a minimal consequence of V, consequently it should be a valid consequence in any reasonable semantics. For an adequate proof theory this would mean that $V \vdash p \vee q$ but neither $V \vdash p$ nor $V \vdash q$ which is not possible in the proof theory for vivid logic presented above. The wellfounded model of V is the empty set, that is, both p and q are considered to be undetermined in this semantics whereas the 3-valued stable model semantics[6] results in three models for V: $\{p, \sim q\}$, $\{\sim p, q\}$ and $\emptyset$. Thus, according to both of these semantics V does not contain any valid information. This seems to indicate that these semantics have serious shortcomings, and cannot be regarded as the generalization and improvement of the natural model semantics for weakly wellfounded VKBs we are looking for.

7.9   Conclusion

I have outlined a partial logic framework for vivid reasoning on the basis of rules overcoming the currently predominant restriction of using only a form of negation-as-failure based on the CWA. Although the suggested formalism is still restricted, it provides considerable expressive power. In particular, it creates the possibility to represent and reason with explicit negative information.

Conditional facts can be viewed as a generalization of definite Horn clauses. It is proposed to use them as program clauses of generalized logic programs. Essentially, this amounts to the introduction of strong negation to logic programming. This proposal fits well with recent developments in logic programming semantics where partial, respectively three-valued, models have become quite popular, yet the possibility of and the need for two different negations has only recently been recognized. Both weak and strong negation have a procedural interpretation. So, conditional facts can be viewed as procedures in a natural way.

Vivid Logic can also serve as a computational framework for nonmonotonic reasoning.[7] The nonmonotonic behaviour of weak negation according to the

---

[6] Cf. [Przymusinski 1990].

[7] See, e.g., [Pearce 1992] where the relationship between Default Logic and Vivid Logic is discussed.

CWA is well-known. But only in interaction with strong negation is sufficient expressiveness obtained.

Chapter 8

Further Topics, Open Problems

8.1 Disjunctive Information

A set of definite KBs forms a disjunctive KB. For instance, a set of VKBs in the sense of chapter 7 is capable of representing and processing disjunctive information.

Consider the KRS

$$\langle 2^{L_{\text{VKB}}}, \vdash, L(-, \sim, \wedge, \vee), \textbf{Upd}, L_{\text{Input}} \rangle$$

called DRV_0, where

1. L_{VKB} is the language of VKBs (see 7.2),

2. derivability from a set Y of VKBs is based on the inference relation of VKBs in the following way:

$$Y \vdash F \stackrel{def}{\Longleftrightarrow} V \vdash F \text{ for all } V \in Y$$

3. $L_{\text{Input}} = \{F \leftarrow G : F \in L(\sim, \wedge, \vee), G \in L(1, -, \sim, \wedge, \vee)\}$,

4. and the update operation is defined for a set of conditional facts $X \in L_{\text{VKB}}$ as input in the following way,

$$\textbf{Upd}(Y, X) := \{V \cup X : V \in Y\}$$

Using this definition, the update operation in the general case can be formulated in terms of the disjunctive normal set:

$$\textbf{Upd}(Y, F \leftarrow G) = \bigcup_{Z \subseteq \text{DNS}(F)} \textbf{Upd}(Y, \{l \leftarrow G : l \in \bigcup Z\})$$

A rule knowledge base X consisting of disjunctive facts and rules of the form
$F \leftarrow G$ where $F \in L(\sim, \wedge, \vee)$ and $G \in L(1, -, \sim, \wedge, \vee)$ can be transformed into
a disjunctive KB of DRV_0 by means of the transformation

$$Y = \mathbf{Upd}(\{\emptyset\}, \bigwedge X)$$

using Conjunction Composition. Notice that the size of Y is exponential in the
number of disjunctive rules $r \in X$, i.e. combinatorial explosion threatens. It
seems, therefore, that the representation of a disjunctive KB as a set of definite KBs is not feasible in practice. It may be more practical not to 'unfold'
all disjunctive rules in advance but only those which are involved in a specific
derivation.

8.2   Conditional Queries

First-degree conditionals $F \rightarrow G$ with definite premise formula, $F \in DefL(\sim$
$, \wedge, \vee)$ and $G \in L(-, \sim, \wedge, \vee)$, do not pose any problem as query formulas. The
inference operation of VKBs can be extended for such queries in a straightfor-
ward way:

$$V \vdash F \rightarrow G \stackrel{def}{\Longleftrightarrow} V \cup K_F \vdash G$$

where $\mathrm{DNS}(F) = \{K_F\}$.

In the general case, where conditionals can be nested and can have indefinite
premises, a set of Harrop programs (see 6.8.4) is needed for representing a KB.
Recall that an arbitrary formula $F \in L(\sim, \wedge, \vee, \rightarrow)$ in $\mathbf{N}$ is equivalent to a
disjunctive normal form based on Harrop programs (see 3.3.4):

$$F \equiv \bigvee_{P \in \mathrm{DNS}(F)} \bigwedge P$$

Let

$$L_H := \{l \leftarrow G : l \in \mathrm{Lit}, G = 1 \text{ or } G = \bigvee \bigwedge Q \text{ such that } Q \subseteq L_H\}$$

denote the set of all Harrop clauses. Then 2^{L_H} is the set of all Harrop programs,
and any set of Harrop programs is called a *Nelson KB*. Such a Nelson KB $\subseteq 2^{L_H}$
allows for inputs $F \in L(\sim, \wedge, \vee, \rightarrow)$ according to the following definition:

$$\mathbf{Upd}(KB, F) := \{P \cup Q : P \in KB, Q \in \mathrm{DNS}(F)\}$$

and

$$KB \vdash F \stackrel{def}{\Longleftrightarrow} P \vdash F \quad \text{for all } P \in KB$$

where $P \vdash F$ is defined as in 6.8.4. In this way, $\mathbf{N}$ can be obtained from the basic VKRS

$$\langle 2^{2^{L_H}}, \vdash, L(\sim, \wedge, \vee, \rightarrow), \mathbf{Upd}, L(\sim, \wedge, \vee, \rightarrow) \rangle$$

which is called *Nelson's KRS*, by

$$F \vdash_N G \quad \text{iff} \quad \mathbf{Upd}(0, F) \vdash G$$

Since the concept of Nelson's KRS is a generalization of Belnap's KRS, one can expect to get similar notions and properties, for instance, an ordering between KBs representing the growth of information.

Question 4 *How can the informational ordering between Nelson KBs be defined ?*

Since it is not clear what contraction means in a program, and it is yet less clear what a weakly negated conditional means, it seems to be difficult to extend Nelson's KRS by the addition of weak negation. One possible extension is to admit weak negation only in the conclusion of a conditional query formula.

The following problems remain open:

Question 5 *How can the deletion of literal information, i.e. $\mathbf{Upd}(P, -l)$, be defined for arbitrary Harrop programs P ?*

Question 6 *What, if anything, does a weakly negated conditional mean ? Or, more precisely, how should $P \vdash -(F \rightarrow G)$ and $\mathbf{Upd}(P, -(F \rightarrow G))$ be defined ?*

8.3 Active Knowledge Bases

A KB is called *active* if it contains action rules of the form

$$Action \leftarrow Condition$$

where *Action* and *Condition*, are respective expressions from an action language L_{Act}, resp. condition language L_{Cond}. An active KB can be modeled as a pair $\langle KB, R_a \rangle$ where KB is a knowledge base from some KRS, and R_a is the set of action rules.

A simple form of an active KRS

$$AK := \langle L_{\text{KB}} \times 2^{L_{\text{Act}} \times L_{\text{Cond}}}, \vdash, L_{\text{Query}}, \mathbf{Upd}_a, L_{\text{Input}} \rangle$$

can be associated with any KRS K in the following way. Let $L_{\text{Cond}} = L_{\text{Query}}$ and $L_{\text{Act}} = L_{\text{Input}}$.[1] Then, the operational semantics of AK is defined as the

---

[1] In the general case, actions will not be restricted to updates but also include private actions, like turning certain switches on and off, and communicative actions, like informing

active closure of KB, denoted by $R_a(KB)$,[2] affecting the update operation of AK:

$$\mathbf{Upd}_a(KB, F) := R_a(\mathbf{Upd}(KB, F))$$

where the active closure $R_a(KB)$ is defined in the same way as the inferential closure $R(KB)$, (see 2.7):

$$r(R_a(KB)) = R_a(KB) \quad \text{for all } r \in R_a,$$

and the application ('firing') of an active rule $r = Input \leftarrow Condition$ is defined as

$$r(KB) = \begin{cases} \mathbf{Upd}_a(KB, Input) & \text{if } KB \vdash Condition \\ KB & \text{otherwise} \end{cases}$$

In an active KB, external updates cause additional (internal) updates by triggering active rules. Since the epistemic component, KB, is always 'saturated' the answering of a query does not involve the active component R_a:

$$\langle KB, R_a \rangle \vdash F \overset{def}{\Longleftrightarrow} KB \vdash F$$

Question 7   *How can conditions of wellfoundedness and consistency be appropriately defined for an active KRS such that $R_a(KB)$ exists uniquely ?*

Example 8.1   *There is a well-established paradigm in the expert system research field which displays exactly the characteristics of nonmonotonic, nonampliative rule-based systems, namely OPS5.[3] OPS5-like systems consist of a set of facts and a set of action rules. The effect of rules are update actions, that is, essentially insertions and deletions. An OPS5 action rule has the form Action $\leftarrow l_1 \wedge l_2 \wedge \ldots \wedge l_n$ where Action is either 'insert a', or 'delete a', and a an atom. Obviously, such rules are special instances of active rules $e \leftarrow F$.*

In [Froidevaux 1992] the following example is discussed:

$$KB = \begin{cases} postgraduate(mike), \ goodworker(mike) \\ goodsearcher(x) \leftarrow postgraduate(x) \wedge goodworker(x) \\ hasoffice(x) \leftarrow goodsearcher(x) \\ poorworker(x) \leftarrow postgraduate(x) \wedge -hasoffice(x) \\ -goodsearcher(x) \leftarrow goodsearcher(x) \wedge poorworker(x) \end{cases}$$

---

other KBs or requesting certain actions from other active KBs. Also, conditions will not only refer to the current epistemic state (KB) but also to the occurence of certain events (recorded, e.g., in a special system component, called 'event buffer').

[2] Notice, however, that the active closure $R_a(X)$ does not need to be unique if it exists, nor does it need to exist at all (see the example below).

[3] See [Forgy 1982].

According to the operational semantics of the execution process of OPS5-based systems this KB does not have a fixpoint (or in other words: there is no active closure of KB). The process of applying successively rules 1, 2, 3 and then 4 can be repeated endless.

Bibliography

[Almukdad & Nelson 1984] A. Almukdad and D. Nelson: Constructible Falsity and Inexact Predicates, *JSL* 49:1 (1984), 231–233.

[Akama 1987] S. Akama: Resolution in Constructivism, *Logique et Analyse* 120 (1987), 385–392.

[Akama 1988a] S. Akama: Constructive Predicate Logic with Strong Negation and Model Theory, *Notre Dame J. of Formal Logic* 29 (1988), 18-27.

[Akama 1988b] S. Akama: On the Proof Method for Constructive Falsity, *Zeitschrift für math. Logik und Grundlagen der Mathematik* 34 (1988), 385–392.

[Akama 1990] S. Akama: Subformula Semantics for Strong Negation Systems, *J. of Philosophical Logic* 19 (1990), 217–226.

[Apt & Bezem 1990] K.R. Apt and M. Bezem: Acyclic Programs, *Proc. ICLP 1990*, MIT Press, 1990.

[Apt, Bol & Klop 1989] K.R. Apt, R.N. Bol and J.W. Klop: On the Safe Termination of Prolog Programs, *Proc. ICLP 1989*, MIT Press, 1989, 353–368.

[Avron 1991] A. Avron: Natural 3-valued Logics – Characterization and Proof Theory, *JSL* 56:1 (1991), 276–294.

[Belnap 1976] N.D. Belnap: How a Computer Should Think, in *Contemporary Aspects of Philosophy*, Proc. Oxford International Symposium 1975, Oriel Press 1976, 30–56.

[Belnap 1977] N.D. Belnap: A Useful Four-valued Logic, in G. Epstein and J.M. Dunn (Eds.), *Modern Uses of Many-valued Logic*, Reidel 1977, 8–37.

[van Benthem 1984] J. van Benthem: The Variety of Consequence, According to Bolzano, *Studia Logica* 44:4 (1984), 389–403.

[Blair & Subrahmanian 1989] H. Blair and V.S. Subrahmanian: Paraconsistent Logic Programming, *Theoretical Computer Science* 68 (1989), 135–154.

[Blamey 1986] S. Blamey: Partial Logic, in D. Gabbay and F. Guenthner (Eds.), *Handbook of Philosophical Logic*, Vol. III, Reidel, Dordrecht, 1986.

[Blau 1978] U. Blau: *Die dreiwertige Logik der Sprache*, de Gruyter, 1978.

[Bolzano 1837] B. Bolzano: *Wissenschaftslehre*, Seidel Buchhandlung, Sulzbach, 1837.

[Bry 1990] F. Bry: Query Evaluation in Recursive Databases: Bottom-up and Top-down Reconciled, *Proc. Int. Conf. Deductive and Object-Oriented Databases (DOOD89)*, North-Holland, 1990, 25–44.

[Clark 1978] K.L. Clark: Negation as Failure, in H. Gallaire and J. Minker (Eds.), *Logic and Databases*, Plenum Press, New York, 1978, 293–322.

[Cleave 1974] J.P. Cleave: The Notion of Logical Consequence in the Logic of Inexact Predicates, *Zeitschrift für mathematische Logik und Grundlagen der Mathematik* 20 (1974), 307–324.

[Codd 1970] E.F. Codd: A Relational Model of Data for Large Shared Data Banks, *Communications of the ACM* 13:6 (1970), 377–387.

[da Costa et al. 1990] N.C.A. da Costa, L.J. Henschen, J.J. Lu and V.S. Subrahmanian: Automatic Theorem Proving in Paraconsistent Logics, in M.E. Stickel (Ed.), *Proc. 10th Int. Conf. Automated Deduction (CADE-10)*, Springer LNAI 449 (1990).

[DiPaola 1969] R.A. DiPaola: The recursive unsolvability of the decision problem for the class of definite formulas, *J. of the ACM* 16:2 (1969), 324 ff.

[Doherty 1991] P. Doherty: NML3 – A Non-Monotonic Formalism with Explicit Defaults, Dissertation, Linköping University, 1991.

[Etherington et al. 1989] D. Etherington, A. Borgida, R. Brachman, and H. Kautz: Vivid Knowledge and Tractable Reasoning: Preliminary Report, *Proc. IJCAI-89*, Morgan Kaufmann, 1989, 1146–1152.

[Fenstad et al. 1987] J.E. Fenstad, P.-K. Halvorsen, T. Langholm and J. van Benthem: *Situations, Language and Logic*, vol. 34 of *Studies in Linguistics and Philosophy*, Reidel, Dordrecht, 1987.

[Fine 1985] K. Fine: *Reasoning with Arbitrary Objects*, Basil Blackwell, 1985.

[Fitch 1948] F.B. Fitch: An Extension of Basic Logic, *JSL* 13 (1948), 95–106.

[Fitch 1952] F.B. Fitch: *Symbolic Logic*, Ronald Press, New York 1952.

[Fitting 1986] M. Fitting: A Kripke-Kleene Semantics for Logic Programs, *JLP* 1986:3, 75–88.

[Forgy 1982] C. Forgy: Rete – A Fast Algorithm for the Many Pattern / Many Object Pattern Match Problem, *AI* 19, 17–37.

[Froidevaux 1992] C. Froidevaux: Default Logic for Action Rule-Based Systems, *Proc. of ECAI-92*, Wiley, 1992, 413–417.

[Fuhrmann 1992] A. Fuhrmann: Subtraction Revisited, preliminary report, Universität Konstanz, 1992.

[Gabbay 1985] D. Gabbay: Theoretical Foundations for Nonmonotonic Reasoning in Expert Systems, in K.R. Apt (Ed.), *Proc. NATO Advanced Study Institute on Logics and Models of Concurrent Systems*, Springer Verlag, 1985, 439–457.

[Gabbay & Sergot 1986] D. Gabbay and M.J. Sergot: Negation as Inconsistency, *JLP* 1986:1, 1–35.

[Gärdenfors 1984] P. Gärdenfors: The Dynamics of Belief as a Basis for Logic, *British J. Philosophy of Science* 35 (1984), 1–10.

[Gärdenfors 1988] P. Gärdenfors: *Knowledge in Flux*, MIT Press, Cambridge, 1988.

[Grzegorczyk 1964] A. Grzegorczyk: A Philosophically Plausible Formal Interpretation of Intuitionistic Logic, *Indagationes Mathematicae* 26 (1964), 596–601.

[Van Gelder, Ross & Schlipf 1988] A. van Gelder, K. Ross and J.S. Schlipf: Unfounded Sets and Well-Founded Semantics for General Logic Programs, *Proc. 7th ACM Symp. Principle of Database Systems*, 1988, 221–230.

[Van Gelder & Topor 1991] A. van Gelder and R.W. Topor: Safety and Translation of Relational Calculus Queries, *ACM Transactions on Database Systems* 16:2 (1991), 235–278.

[Gelfond & Lifschitz 1988] M. Gelfond and V. Lifschitz: The Stable Model Semantics for Logic Programming, *Proc. ICLP 1988*, MIT Press, 1988.

[Gelfond & Lifschitz 1990] M. Gelfond and V. Lifschitz: Logic Programs with Classical Negation, *Proc. ICLP 1990*, MIT Press, 1990.

[Gelfond & Lifschitz 1991] M. Gelfond and V. Lifschitz: Classical Negation in Logic Programs and Disjunctive Databases, *J. New Generation Computing*, 9 (1991), 365–385.

[Ginsberg 1986] M.L. Ginsberg: Multivalued Logics, *Proc. of AAAI'86*, 1986, 243–247.

[Hallnäs & Schroeder-Heister 1987] L. Hallnäs and P. Schroeder-Heister: A Proof-Theoretic Approach to Logic Programming I: Generalized Horn Clauses, SICS Research Report, 1987; also in *J. of Logic and Computation*.

[Herre 1988] H. Herre: Negation and Constructivity in Logic Programming, *J. of New Generation Computer Systems* 1 (1988), 295–305.

[Herre 1991] H. Herre: Foundations of Clause Logic Programming, manuscript, read at CSL'91, 1991.

[Herre & Pearce 1992] H. Herre and D. Pearce: Disjunctive Logic Programming, Constructivity and Strong Negation, in D. Pearce and G. Wagner (Eds.), *Proc. of 3rd European Workshop on Logics in AI (JELIA'92)*, Springer LNAI 633 (1992).

[Johannson 1937] I. Johannson: Der Minimalkalkül - ein reduzierter intuitionistischer Formalismus, *Compositio Mathematica* 4 (1937), 119–136.

[Kleene 1952] S. Kleene: *Introduction to Metamathematics*, Van Nostrand, Princeton, 1952.

[Körner 1966] S. Körner: *Experience and theory*, Kegan Paul, London 1966.

[Kowalski & Sadri 1990] R.A. Kowalski and F. Sadri: Logic Programs with Exceptions, *Proc. ICLP 1990*, MIT Press, 1990, 598–613.

[Kutschera 1984] F. Kutschera: Eine Logik vager Sätze, *Archiv für mathematische Logik und Grundlagenforschung* 24 (1984), 101–118.

[Kutschera 1985] F. Kutschera: *Der Satz vom ausgeschlossenen Dritten*, de Gruyter, 1985.

[Langholm 1988] T. Langholm: *Partiality, Truth and Persistence*, CSLI Lecture Notes 15, University of Chicago Press, 1988.

[Levesque 1984a] H.J. Levesque: Foundations of a Functional Approach to Knowledge Representation, *AI* 23:2, 1984, 155–212.

[Levesque 1984b] H.J. Levesque: The Logic of Incomplete Databases, in M.L. Brodie and J.W. Schmidt (Eds.), *On Conceptual Modelling: Perspectives from AI, Databases and Programming Languages*, Springer-Verlag, 1984.

[Levesque 1986] H.J. Levesque: Making Believers out of Computers, *AI* 30 (1986), 81-107.

[Levesque 1988] H.J. Levesque: Logic and the Complexitiy of Reasoning, *J. Philosophical Logic* 17 (1988), 355–389.

[Lifschitz & Woo 1992] V. Lifschitz and T.Y.C. Woo: Answer Sets in General Nonmonotonic Reasoning (Preliminary Report), *Proc. KR'92*, Morgan Kaufmann, 1992, 603–614.

[Lloyd & Topor 1984] J.W. Lloyd and R.W. Topor: Making Prolog More Expressive, *J. Logic Programming* 1984:3, 225–240.

[Lukasiewicz 1920] J. Lukasiewicz: On 3-valued logic, in S. McCall (Ed.): *Polish Logic*, Oxford University Press, 1967.

[Makinson 1989] D. Makinson: General Theory of Cumulative Inference, in M. Reinfrank et al. (Eds.), *Proc. 2nd Int. Workshop on Nonmonotonic Reasoning*, Springer LNAI 346 (1989), 1–18.

[Marek & Truszczynski 1991] W. Marek and M. Truszczynski: Autoepistemic Logic, *Journal of the ACM* 38 (1991), 588–619.

[Markov 1950] A.A. Markov: A Constructive Logic (in Russian), *Uspeki Mathematiceki Nauk* 5 (1950), 187–188.

[Miller 1989] D. Miller: A Logical Analysis of Modules in Logic Programming, *J. Logic Programming* 1989, 79–108.

[Moore 1982] R.C. Moore: The Role of Logic in Knowledge Representation and Commonsense Reasoning, *Proc. AAAI-82*, 1982, 428–433.

[Nelson 1949] D. Nelson: Constructible falsity, *JSL* 14 (1949), 16–26.

[Niemelä & Rintanen 1992] I. Niemelä and J. Rintanen: On the Impact of Stratification on the Complexity of Nonmonotonic Reasoning, *Proc. KR'92*, Morgan Kaufmann, 1992, 657–638.

[Nute 1990] D. Nute: Editorial, *J. Logic and Computation* 1:2 (1990), 155–158.

[Patel-Schneider 1990] P.F. Patel-Schneider: A Decidable First-Order Logic for Knowledge Representation, *J. Automated Reasoning* 6 (1990), 361–388.

[Pearce 1991] D. Pearce: n Reasons for Choosing N, LWI Technical Report 14/1991, Freie Universität Berlin. Revised version to appear in *J. Logic, Language and Information*.

[Pearce 1992] D. Pearce: Default Logic and Constructive Logic, *Proc. ECAI-92*, Wiley, 1992.

[Pearce & Wagner 1989] D. Pearce and G. Wagner: Reasoning with Negative Information I – Strong Negation in Logic Programs, LWI Technical Report, Freie Universität Berlin, 1989. Also in L. Haaparanta, M. Kusch and I. Niiniluoto (Eds.), *Language, Knowledge, and Intentionality*, Acta Philosophica Fennica 49, 1990.

[Pearce & Wagner 1990] D. Pearce and G. Wagner: Logic Programming with Strong Negation, in P. Schroeder-Heister (Ed.), *Proc. Workshop on Extensions of Logic Programming*, Springer LNAI 475 (1990).

[Pereira & Alferes 1992] L.M. Pereira and J.J. Alferes: Wellfounded Semantics for Logic Programs with Explicit Negation, *Proc. ECAI'92*, Wiley, 1992.

[Pereira, Alferes & Aparicio 1992] L.M. Pereira, J.J. Alferes and J.N. Aparicio: Contradiction Removal Semantics with Explicit Negation, *Proc. Applied Logic Conference Amsterdam 1992*.

[Prawitz 1965] D. Prawitz: *Natural Deduction*, Almqvist and Wiksell, Stockholm 1965.

[Przymusinski 1988] T.C. Przymusinski: On the Declarative Semantics of Deductive Databases and Logic Programs, in J. Minker (Ed.), *Foundations of Deductive Databases and Logic Programming*, Morgan Kaufmann, 1988, 193–216.

[Przymusinski 1989a] T.C. Przymusinski: Non-Monotonic Formalisms and Logic Programming, *Proc. ICLP 1989*, MIT Press, 1989.

[Przymusinski 1989b] T.C. Przymusinski: Every Logic Program has a Natural Stratification and an Iterated Least Fixed Point Model, *Proc. 8th ACM Symp. on Principles of Database Systems*, 1989, 11–21.

[Przymusinski 1990] T.C. Przymusinski: Well-founded Semantics Coincides with Three-valued Stable Semantics, *Fundamenta Informaticae* XIII (1990), 445–463.

[Przymusinska & Przymusinski 1990] H. Przymusinska and T.C. Przymusinski: Weakly Stratified Logic Programs, *Fundamenta Informaticae* XIII (1990), 51–65.

[Quantz & Kindermann 1990] J. Quantz and C. Kindermann: Implementation of the BACK-System 4, KIT-Report 78, Technische Universität Berlin, 1990.

[Rajasekar, Lobo & Minker 1989] A. Rajasekar, J. Lobo and J. Minker: Weak Generalized Closed World Assumption, *J. Automated Reasoning* 5 (1989), 293–307.

[Rasiowa 1974] H. Rasiowa, *An Algebraic Approach to Non-classical Logics*, North-Holland, 1974.

[Rautenberg 1979] W. Rautenberg: *Klassische und nichtklassische Aussagenlogik*, Vieweg, 1979.

[Reiter 1978] R. Reiter: On Closed-World Databases, in J. Minker and H. Gallaire (Eds.): *Logic and Databases*, Plenum Press, 1978.

[Reiter 1990] R. Reiter: On Asking What a Database Knows, in J. Lloyd (Ed.), *Computational Logic*, Proceedings, Springer-Verlag, 1990, 96–113.

[Rescher 1976] N. Rescher: *Plausible Reasoning*, Van Gorcum, Assen/Amsterdam, 1976

[Sakama 1992] Ch. Sakama: Extended Well-Founded Semantics for Paraconsistent Logic Programs, *Proc. Int. Conf. on Fifth Generation Computer Systems 1992*, ICOT, 1992, 592–599

[Sergot, Sadri, Kowalski et al. 1986] M.J. Sergot, F. Sadri, R.A. Kowalski, F. Kriwaczek, P. Hammond and H.T. Cory: The British Nationality Act as a Logic Program, *Communications of the ACM* 29:5 (1986), 370–386.

[Sheperdson 1988] J.C. Sheperdson: Introduction to the Theory of Logic Programming, *Proc. of Logic Colloquium 1986*, North-Holland, 1988.

[Tan 1992] Y.H. Tan: Non-Monotonic Reasoning: Logical Architecture and Philosophical Applications, Dissertation, Free University Amsterdam, 1992.

[Thijse 1992] E. Thijse: Partial Logic and Knowledge Representation, Dissertation, Katholieke Universiteit Brabant, Eburon Publishers, Delft, 1992.

[Thomason 1969] R.H. Thomason: A Semantical Study of Constructible Falsity, *Zeitschrift für math. Logik und Grundlagen der Mathematik* 15 (1969), 247–257.

[Toulmin 1956] S. Toulmin: *The uses of argument*, Cambridge University Press, Cambridge, 1956

[Urbas 1990] I. Urbas: Paraconsistency, *Studies in Soviet Thought* 39 (1990), Kluwer, 343–354.

[Vardi 1981] M. Vardi: The decision problem for database dependencies, *Information Processing Letter* 12:5 (1981), 251–254.

[Veltman 1990] F. Veltman: Defaults in Update Semantics, in H. Kamp (Ed.), *Conditionals, Defaults and Belief Revision*, Edinburgh, Dyana deliverable R2.5.A, 1990.

[Wagner 1989] G. Wagner: Algebraic Semantics of Propositional Logic Programs, LWI Technical Report 7/1989, Freie Universität Berlin. Also in D. Pearce and H. Wansing (Eds.), *Nonclassical Logic and Information Processing*, Springer LNAI 619 (1992).

[Wagner 1990] G. Wagner: The two Sources of Nonmonotonicity in Vivid Logic: Inconsistency Handling and Weak Falsity, in G. Brewka and H. Freitag (Eds.), *Proc. GMD Workshop on Nonmonotonic Reasoning 1989*, Gesellschaft für Mathematik und Datenverarbeitung, Bonn - St. Augustin, 1990

[Wagner 1991a] G. Wagner: A Database Needs Two Kinds of Negation, *Proc. 3rd Int. Symp. on Mathematical Fundamentals of Database and Knowledge Base Systems MFDBS-91*, Springer LNCS 495 (1991), 357–371.

[Wagner 1991b] G. Wagner: Logic Programming with Strong Negation and Inexact Predicates, *J. Logic and Computation* 1:6 (1991), 835–859.

[Wagner 1991c] G. Wagner: Ex contradictione nihil sequitur, *Proc. IJCAI'91*, Morgan Kaufmann, 1991, 538–543.

[Wansing 1992] H. Wansing: The Logic of Information Structures, Dissertation, Free University Berlin, 1992, to appear as Springer LNAI (1993).

[Wittgenstein 1956] L. Wittgenstein: *Remarks on the Foundations of Mathematics*, MacMillan, New York, 1956.

Appendix A

An Interpreter for VL in Prolog

The following Prolog program works as an inference engine processing queries to a well-founded VKB. Conditional facts have to be entered to the Prolog database in the form `1 <- F` where `1 = a | ~a` and `F = 1 | 1 | F & F | F v F | ~F | -F`.

```
?- op( 100, fy, ~).        /*  strong negation                */
?- op( 100, fy, -).        /*  weak negation                  */
?- op( 110, xfy, &).       /*  conjunction                    */
?- op( 120, xfy, v).       /*  disjunction                    */
?- op( 130, xfy, <-).      /*  conditional fact separator     */

prove(1).
prove( F v G)    :-  !, ( prove( F); prove( G)).
prove( F & G)    :-  !, ( prove( F), prove( G)).
prove(~(F v G))  :-  !, ( prove(~F), prove(~G)).
prove(~(F & G))  :-  !, ( prove(~F); prove(~G)).
prove(-(F v G))  :-  !, ( prove(-F), prove(-G)).
prove(-(F & G))  :-  !, ( prove(-F); prove(-G)).
prove(-~(F v G)) :-  !, ( prove(-~F); prove(-~G)).
prove(-~(F & G)) :-  !, ( prove(-~F), prove(-~G)).
prove(~~F)  :-  !, prove( F).
prove(--F)  :-  !, prove( F).
prove(~-F)  :-  !, prove( F).
prove(-~-F) :-  !, prove(-F).
prove(~~-F) :-  !, prove(-F).
prove(-L)   :-  !, not prove( L).
```

```
/* Closed-World Assumption           */

prove(~A) :-  !, ( ~A <- F, prove( F);
               A =.. [P|Args], cwa( PT, _),
               member( P, PT), not prove( A)).

prove( A) :-  !, ( A <- F, prove( F);
               A =.. [P|Args], cwa( _, NT),
               member( P, NT), not prove(~A)).
```

Index

Springer-Verlag
and the Environment

We at Springer-Verlag firmly believe that an international science publisher has a special obligation to the environment, and our corporate policies consistently reflect this conviction.

We also expect our business partners – paper mills, printers, packaging manufacturers, etc. – to commit themselves to using environmentally friendly materials and production processes.

The paper in this book is made from low- or no-chlorine pulp and is acid free, in conformance with international standards for paper permanency.

Printing: Weihert-Druck GmbH, Darmstadt
Binding: Buchbinderei Schäffer, Grünstadt

Lecture Notes in Artificial Intelligence (LNAI)

Lecture Notes in Computer Science